In Praise of Beer

In Praise of Beer

CHARLES BAMFORTH

OXFORD
UNIVERSITY PRESS

Oxford University Press is a department of the University of Oxford. It furthers
the University's objective of excellence in research, scholarship, and education
by publishing worldwide. Oxford is a registered trade mark of Oxford University
Press in the UK and certain other countries.

Published in the United States of America by Oxford University Press
198 Madison Avenue, New York, NY 10016, United States of America.

Library of Congress Cataloging-in-Publication Data
Names: Bamforth, Charles, 1952– author.
Title: In praise of beer / Charles Bamforth
Description: New York, NY : Oxford University Press, [2020] |
Includes bibliographical references and index.
Identifiers: LCCN 2019035556 (print) | LCCN 2019035557 (ebook) |
ISBN 9780190845957 (hardback) | ISBN 9780190845971 (epub) |
ISBN 9780190845964 (updf)
Subjects: LCSH: Beer.
Classification: LCC TP577 .B3445 2020 (print) | LCC TP577 (ebook) |
DDC 641.2/3—dc23
LC record available at https://lccn.loc.gov/2019035556
LC ebook record available at https://lccn.loc.gov/2019035557

1 3 5 7 9 8 6 4 2

Printed by Sheridan Books, Inc., United States of America

For John Hudson, Tony Portno, Gus Guthrie, Bernard Atkinson, Graham Stewart, Doug Muhleman, and Ken Grossman, the people who did more than anyone to shape my journey in beer.

Contents

Preface

There is a supposed Chinese curse that says, "May you live in interesting times." There is no doubt whatsoever that, when it comes to beer, these most certainly are extremely interesting times. In China itself the brewing of beer has accelerated at an astonishing rate in the past couple of decades. Elsewhere, in a huge range of countries but perhaps best typified by the likes of the United States, United Kingdom, and Australia, there is a burgeoning so-called craft sector, with a vast growth in the number of brewing companies. Here in California, as elsewhere, there is no end to the ingenuity of these brewers, who are forever pushing the boundaries in terms of styles, ingredients, and presentation of products that can either delight, disturb, or distress the drinker, depending on perceptions, preferences, and preconceived biases. No matter, the reality is that the beer world is emerging and exciting. There is an ongoing need for new brewers who are well informed and capable—for which folks like me, whose day job has been to make a living out of teaching, are inordinately grateful. Equally, there seems to be a growing thirst from customers, not only for the beers themselves but also for an understanding of what they are drinking.

Most of the books I have written over a quarter of a century have primarily been targeted at the producers of beer. This one, however, has been penned largely with the customer in mind, although I hope that won't stop those employed by brewing companies from reading it, because they sure need to know what I am preaching to the customer. Customers are becoming more knowledgeable and, therefore, more choosy and, yes, demanding. It's a good thing, provided that they speak from a position of genuine understanding. I hope that this volume will help.

Gratitude

There was no master plan. My journey into and through the brewing industry was a case of "It seemed the best thing to do at the time," although I am grateful to a good many people for nudging me (gently or sometimes with more vigor) into the next step.

It started near Wigan in Lancashire, England, when my elder brother (by 18 months) John was my first chemistry teacher. I understood things like valence, balancing equations, and the symbols for the elements at the age of nine or 10. We had a rather well set-up laboratory at home and would have chemicals delivered by the Gallenkamp company, just one of their stops on the rounds that took them to colleges, schools, and hospitals. It was also John who put pressure on me when, at the age of 15, I had to decide the subjects for my advanced-level studies at Up Holland Grammar School. Students had to take three A-level courses, and, on the basis of my lackluster O-level performance in eight subjects, it was a moot point whether I was best suited to the arts or the sciences. My heart told me English, geography, history. My brother told me (and my teacher mother) chemistry, biology, physics. (Mathematics was agreed to be the one no-no.) "You will stand to more readily find a career with those sciences than with the arty topics," John insisted.

Things started well enough in the A levels, but things started to decline as I was distracted by other things, mostly of a sporting variety but also the joys of the local hostelries, frequented on Friday evenings from an age rather less than the legal 18. No more than a couple of pints, of course. In truth, though, I was not the best of students, mostly my own fault but not helped by teachers such as Mrs. Brown, whose initial words to me when I came first into her Biology class were, "I don't know what you are doing here, Bamforth." (Years later, when I was elected a fellow of the Royal Institute of Biology, I was tempted to try to find her.) Luckily, though, there was another, rather kinder teacher and the head of Biology, "Spike" Jones. And it was his teaching in the limited coverage of matters biochemical that I found the most fascinating component of all my A-level studies.

Somehow I got half decent grades in my examinations and succeeded in gaining entry to my first choice establishment, the University of Hull, to read biochemistry. It was at this point that I really buckled down to my studies, devouring the subject matter. I also discovered that Hull is a beer drinker's paradise, and Friday and Saturday evenings saw me very much enjoying a plethora of public houses.

Having gained a good-class BSc, I stayed in Hull to do my PhD with the lovely Peter Large, a super scientist and beer bon viveur. It was in this period from 1973 to 1976 that I established my love affair with enzymology and also the fascination for research.

Thence to a postdoctoral fellowship at the University of Sheffield with the brilliant Rod Quayle, FRS. The first year saw me delving ever deeper into the enzymes of a certain type of bacteria, work that carried on in the second year, but in that latter period I spent considerable time scrutinizing the back pages of magazines like *Nature* and the *New Scientist* in search of job openings. The first to catch my eye was one for an enzymologist at the Brewing Research Foundation (BRF). My, my: capturing two of my four loves, beer and enzymes, in one fell swoop (the other two loves, of course, being my new wife and Wolverhampton Wanderers Football Club).

BRF (see chapter 2) was a remarkable organization, funded by a research tax on UK beer that went to pay for a battery of excellent scientists with diverse specializations to do fundamental pre-competitive work on beer and brewing to the advantage of all. Things went well—although I hadn't been at Nutfield terribly long before I questioned whether my scientific abilities were being properly used: should I not be in a medical field, trying to unravel the mysteries of disease? When I told Acting Director John Hudson of my doubts, he looked at me plainly and asked me whether people enjoyed beer. I replied that they did. "So you don't believe that it is a noble calling to do research to ensure that they contribute to gain happiness?" I had no reply. "So b***er off, and get on with it." I have strived to do that ever since.

The biggest paymasters to BRF were the ones who brewed the most beer in the nation, as the tax was in proportion to barrels of beer sold. Thus Bass was the major funder and, like the other members, it viewed BRF as a recruitment agency. Hence I was spirited away to Bass and soon became the research manager in Burton-on-Trent, thanks to the determination of Tony Portno. It was a subsequent boss, Gus Guthrie, however, who was most adamant that, having been in the ivory tower for five years at BRF and another five at Burton, I should be seen to get (and benefit from) experience at the coalface, so to speak. Hence I was dispatched to the Preston Brook Brewery as the quality assurance manager. Great experience. I was nowhere near as personally satisfied there as I had been in research, but, my word, I learned lot about what happened in "the real world," much of it not to be discovered easily in a textbook.

After two and a bit years I accepted the invitation to return to BRF as director of research, reporting to the unusual but effective Director-General Bernard Atkinson. Soon afterward my close friend Graham Stewart, erstwhile technical guru at Labatt in London, Ontario, but now professor in Heriot-Watt University in Edinburgh, Scotland, invited me to become visiting professor of brewing (in the United States it would be called adjunct professor) in that famous brewing education establishment.

Thus in 1998 when the University of California, Davis, came looking for the first Anheuser-Busch Endowed Professor, insisting on someone with a strong record in relevant research, teaching, and practical experience, it certainly did seem to be rather up my street. I arrived in 1999, grateful to Doug Muhleman from the historic US brewing company, who was not only the driving force behind establishing the professorship but also instrumental in supporting me as I shifted the facilities from a dismal condition not enhanced since the first pilot brewery was donated to the campus by the Lucky Lager Brewing Company in 1958 to a place that we could justifiably be proud of.

Innumerable other individuals and companies supported the cause through my tenure, most notable of all being Ken Grossman of Sierra Nevada Brewing Company.

And since 1972 my partner in all of this has been my wife of 42 years, Diane, without whom it would not have happened.

In Praise of Beer

1

Perceptions

I cheerily made my way beneath west London's soggy blackness to the welcoming warmth of the King William (Fig. 1.1), a quite regular haunt on my visits to the mother country and very much a friendly confine. "A pint of IPA please."

The lass behind the bar had difficulties ("He's having problems with the cork"—by which I assume she meant the tapping of the barrel in the cellar by mine host), but eventually I got a pint of Greene King's finest (Fig. 1.2). It was on reasonable form and I settled down to Simon Brett's *A Reconstructed Corpse*, with an eye on the television, where Arsenal was taking a quick two-goal lead over Everton. The surroundings were familiar and convivial. The room was bright and toasty. God was in his or her heaven and all was well.

There is something of an unpredictability about cask beer. I often tell my classes that when I mimic (air guitar style) the pulling of a pint of cask-conditioned ale, my mouth starts to salivate. It is the beery equivalent of Pavlov's dogs. When the event occurs for real in one of England's fine hostelries, it must be said that there is a certain capriciousness therein. I would say that on 70% of occasions (tonight included) the pouring was acceptable if unspectacular. Sometimes it is a drink to die for—I would even suggest that it is the most drinkable beer that one can get. Unfortunately, there are occasions when it is an experience that makes you almost feel that dying would be the kinder way out.

In due course I did the polite thing, took my empty glass back to the bar, and made myself comfortable with a visit to the compact but functional Gents before strolling through the raindrops across the road to another habitual haunt, the Achari (Fig. 1.3). I'm sufficiently well known to warrant a handshake from the welcoming waiters, an ushering to my favorite table, and the ready understanding that the evening should begin with a pint of Carlsberg Extra (Fig. 1.4) and two papadums with the trimmings (which for me will be lashings of onions bathed in yogurt sauce).

The beer is cold, bright, and fizzy, presented to perfection in a glass that nucleates the bubbles. The beverage is an ideal accompaniment to the

Fig. 1.1. A favorite pub. Photo by the author.

Fig. 1.2. A favorite beer. Photo by the author.

Fig. 1.3. A favorite restaurant. Photo by the author.

Fig. 1.4. Another fine beer. Photo by the author.

appetizer and the subsequent delights of prawn puri, chicken pasanda, and pilau rice. The Carlsberg Extra was refreshing, drinkable (perhaps too much—though I held back to two pints on this occasion), and tasted entirely as I predicted that it would taste. After all, I have been savoring Indian food quenched with the Danish company's finest for nearly half a century. I could of course have ordered Cobra or Kingfisher (was there ever a more delightful logo for a beer?), and often I do in the vast number of restaurants serving sub-continental cuisine that I frequent. The reality is that it is these cold, sound-bodied but non-extreme, pale lager beers that are a magnificent accompaniment to Indian cuisine. The Greene King IPA would not hack it alongside a vindaloo and an onion bhaji. It knows its place: alongside a ploughman's lunch, a Melton Mowbray pork pie, or some black pudding.

And how so? Is it a genuine matter of true food-drink pairing realities? Or is it rather more a psychological matter, with a deeply ingrained prejudice and mindset etched into my gray matter over years of developing certainty of what is right and what is wrong?

Carlsberg: one of the world's top four brewing companies, who would advertise their beer on a platform of "probably the best lager in the world." Irritating and subjective—but definitely memorable and therefore, I presume, much admired in the trendy and insincere worlds of marketing and advertising.

Greene King: a long-standing and very traditional brewer of classic cask-conditioned ales and thereby a darling of the Campaign for Real Ale (CAMRA). The CAMRA remit until very recently was to champion only beer ("real ale") that was produced according to very restricted production techniques and packaged into casks. They would fundamentally decry anything else (e.g., Carlsberg Extra) as being indistinguishable from insipid piss.

For me, on a dank February evening in the western suburbs of London, both beers were totally fit for purpose. I would never have dreamed of asking in a time-honored low-beamed pub for a pint of lager, though many people (actually the larger proportion) do, and it really does not make them less worthy than me. And no way would I have ordered in the Indian restaurant a pint of ale (although I'm pretty sure that they didn't have any), and much less would I have ordered a wine, despite the fact that this restaurant proudly advertises a special wine (their sales pitch) to accompany Indian food. Some people, however, are prepared to quaff chardonnay alongside their curry—and that, too, is okay. I suppose.

Which brings me by way of illustration to the latter-day reality of a world suffused with bigotry, intolerance, and an inability to respect the contrary opinions of another. I am of course talking about pretty much anything in these extreme days, not least politics. Let us stick to beer, though. It's safer. I think.

For some it simply boils down to a matter of size, big being decidedly un-beautiful. Devotees of "craft" beer perceive it as ludicrous to think that the mighty conglomerates are capable of brewing anything close to being drink-able. A gently flavored North American lager might as well be something that has been filtered through a kidney. Those who must drink such beer, if such they be, must surely have no true appreciation of what a decent drink must comprise—to the minds of these critics a product with enough aroma and taste to smack your senses into submission. Some moron coined the term "yellow fizzy liquid." Others speak unknowingly of "industrial beer."

The reality, of course, is that many more gallons of these types of beer are consumed in the United States than of those styles possessed of more com-plexity and flavor. As I write, the top ten beers in terms of sales are Bud Light, Coors Light, Budweiser, Miller Lite, Michelob Ultra Light, Natural Light, Busch Light, Busch, Miller High Life, and Keystone Light. Now of course there are myriad factors at play: these beers are made by two vast compa-nies that have gigantic sway on the distribution chain, immense resources for advertising, and an inherent capacity, founded on sheer volume, to pitch some products at ridiculously competitive rates. But if these products really were so undrinkable, why would they be quaffed in such vast amounts? And no, it is not because their drinkers are ignorant and unknowing. It is because these are the beers that these customers are used to drinking, want to drink, and they have become the brands to which these drinkers believe they owe loyalty.

Are they in my refrigerator? No. Why? Because my personal preferences are not for the style in question—although there is many a drinking occasion when a gently nuanced beer on this list is mightily refreshing. Think of the furnace that is a summer's afternoon here in the Central Valley of California. However, you won't get me to rubbish these people (either the brewers or the drinkers) even if a disproportionately large number of the consumers choose to quaff these brews straight out of the bottle or can, which I just don't get, as much as anything because you don't know what has crawled across the top of the container as it made its way to you.

Many people say, well these are cheap because they are made with down-priced raw materials like rice and corn. Let's consider one example of this: Budweiser, which first saw the light of day in 1876. (Isn't this brand—and its mode of production—traditional by now, more than 140 years later?) Contrary to popular prejudice, it is not made just from rice. The grist is 70% malted barley and 30% rice. The latter comes from Northern California and Arkansas, and huge effort is devoted to making it fit for purpose, with extremely careful milling to ensure that there is no remaining germ, the fats from which would make it turn rancid rapidly. Historically, Anheuser-Busch devoted vast attention to the transport conditions to ensure that the material got to the brewery and was used as rapidly as possible. And to use it adds an additional expensive process stage in the brewhouse. Rice (and corn for that matter) is not primarily used on the grounds of cost, but rather to lighten the color and flavor. The outside layer of barley contains materials that give grainy and astringent (drying) character. Using rice tones these characters (which can be disagreeable) down.

But the prejudice is that these brewers must be bad people for using the likes of rice and corn. This puzzles me, because rice and corn are, like barley and wheat, merely types of cereal grain and, as such, just different packages of starch. It seems to the uninformed that wheat is okay (think *Hefeweizen*) and oats, too (e.g., oatmeal stout). But rice and corn? Get out of here. They might dwell on the fact that Sir Walter Raleigh (Fig. 1.5), who made the first beer by a European in North America, used a grist of corn. Bad person?

Consider, though, the Trappist monk. He boosts the sugar content of his brews using candi sugar, which is derived from sugar beet. Well that must be okay, says the beer bigot. God is on board with that one. And what about Wynkoop from Denver, with their Rocky Mountain oysters or Green Man in Wellington, New Zealand, with their beer suffused with stag semen? Well they are innovative, folks say.

As one of my favorite footballers (aka soccer players, if you must), Jimmy Greaves, used to say: "It's a funny old game." And that is what beer and brewing is these days: a game wherein players on the same team, namely the brewing team, fight one another, while the opposition, the winemakers, quietly get on with pulling in a common direction.

In traditional beer domains, such as Belgium, the Czech Republic, Germany, Ireland, and the United Kingdom, per capita consumption of beer plummets, while that of wine is significantly higher now than it was half a century ago, although there has been somewhat of a downturn in overall

Fig. 1.5. Sir Walter Raleigh. National Portrait Gallery. Public domain.

alcohol consumption in the last few years. The same situation obtains in the United States, although the overall decline in beer volumes is not as precipitous as elsewhere. The reasons are complex, but they surely include factors such as wine being perceived as healthier, more socially acceptable in the circles that the younger drinker aspires to, and more suited to the dinner table. I would contend that beer is a better beverage in all of these and in other respects and will emphasize this throughout this book.

Much needs to be done to recover the high ground for beer. I do not believe that infighting helps.

So we now have the notion of "craft brewing." On the one hand it is something that has led to a much-welcomed increase in interest in beer across the globe. However, it too, with terms like "industrial brewing," led to a confusion in the mind of ill-informed customers about how their beer is made. The sense is that a craft brewer toils in a dimly lit facility, passionately and painstakingly bringing forth rich goodness from lovingly selected and very local raw materials, every drop testimony to devotion and concern. On the other hand, the big guys are portrayed as brewing in vast factories, concocting spurious blends by some underhand chemical processing, preserving the beers with goodness knows what unsavory poison and laughing all the way to the bank.

The reality is very different. It is true that there are a great many tiny brewing operations that are very much doing wonders through the sheer dedication to their craft. Equally there are some people getting into brewing who absolutely shouldn't, because they know precious little about the science and technology of brewing (for it is a science and a technology), and they have eyes only for the day when some rich conglomerate will pay many millions to send them off onto a life of luxury.

And there are no chemistry sets in the big brewing companies. The truth is that the better ones have astonishing commitment to selecting the choicest raw materials, brewing them with time-honored processes, albeit protocols that are finely controlled using sensors and other state-of-the-art strategies, and striving to put beer into package that has got the best possible shelf life and which will delight customers by meeting their expectations every time. The large companies employ (and pay well) highly trained and dedicated people who are very much craftspeople.

In the technical production of beer there simply cannot be any correlation between size and the ideal of brewing excellence. The biggest companies are capable of making any beer (and I mean any beer) astonishingly well. There's many a small company that can do that—but only those who understand the essentials of the process. Those who don't, make swill.

Nevertheless, attempts are made to define a craft brewery on parameters that very much include size. Notably we have the definitions from the Brewers Association, the Boulder-based body that looks after the "craft" sector. A number of factors enter into their definition, including that another brewing concern can't own more than 25% of their stock. But the key number is six million. A US craft brewing company can brew up to six million barrels per annum, which is an output similar to that from the whole of Ireland.

The reasoning on this is to allow growth, such that some pretty big brewing companies can remain in the club even if they double in size or even more. Notable among these players is the Boston Beer Company. Actually some revised definitions brought another company into the association, namely Yuengling, which first set up shop in 1829 in Pottsville, Pennsylvania (Fig. 1.6). It suddenly went from a company (and a beer) that nobody considered in the context of "craft" to being the biggest operator in the sector. The Bostonian nose was well and truly put out of joint.

Fig. 1.6. The Yuengling brewery in Pottsville, Pennsylvania. Photo courtesy of Jennifer Yuengling.

I weary of the concept of trying to pigeonhole beer quality on the basis of size. Rather I would prefer to consider brewing companies on the basis of their attitudes. Do they care about their employees, their customers, the environment, society in general, and the consistent excellence of their beer? Those things can't readily be linked to volume—although for a company to get truly massive does usually entail a lot of buying of other companies and often spitting out of the unwanted in terms of brewery closures and a resultant struggle for local communities. It was ever thus. The Bass Company that I was proud to be part of grew to being one of the biggest in Europe through some fairly brutal acquisitions and rationalization, as we shall see in chapter 9. By contrast, the even more massive Anheuser-Busch (as it was before it was bought by Inbev) largely grew organically by building newer and newer breweries across the States.

There are too many small brewing companies with bad attitudes. Consider the Atlanta-based brewery whose staff were photographed raising their middle fingers at critical customers. Equally there have been some regrettable postures taken by larger entities. Consider the company that used to put its prospective employees through lie-detector tests. Ponder the gigantic company that operates today on the basis of making its suppliers wait 120 days to get paid. Or the craft brewing companies that marketed their beers by rubbishing the offerings of large domestic and international brewing companies. In particular they would urge people to drink American beer and not that coming into the country in green glass bottles.

There are good and bad attitudes at all sizes of brewing. It has nothing to do with the beer. I get it why a small guy or gal should deplore the big guys, particularly when they masquerade as being smaller than they are with the positioning of some brands. The unknowing perceive them as "craft" when the reality is that they are, as some people choose to call it, "crafty." But I know why they do it, because there are many misinformed drinkers who will automatically infer that something bearing the name of one of the larger concerns must obviously be rubbish. That is a very silly attitude, but it is born through this mutual loathing that has built up in the brewing world. I recall a woman saying to me once how much she enjoyed Blue Moon. I replied that it is a good beer and just shows what a big brewer tends to do, namely make beer consistently well. "What do you mean?" she asked. I explained that Blue Moon was developed by Coors. "Oh, I hate it" she said.

The big companies are decried for buying up the smaller craft breweries, and the sellers are ostracized for cashing in. An example would be Elysian, which used to rejoice in the byline "Corporate Beer Sucks" but which in due course sold to Anheuser Busch Inbev. Few people would surely turn away a multi-million-pound offer. Straight away the beers of those companies are deemed unacceptable—and chances are that people will dismiss them. The word will spread that such and such a beer has been ruined. The reality is that the big companies, with their extensive technical resources, will make those beers more consistently than they were ever brewed before.

Do I totally approve? No, because these purchased entities now have access to the technical and purchasing might of the corporate conglomerates. They have an unfair advantage over those that remain small (whatever small actually means). And I don't like the hold on purchasing raw materials that size enables and the reality that beer distributors will jump far higher for the larger entities, who therefore have much more ability to get their drinks into the public eye and onto the shelves.

Surely, overall though, the notion of craftspersonship is not about size or ownership. It is about attitude. Companies like Sierra Nevada in the United States and Coopers in Australia are marvelous family-owned entities that have refused any overtures from larger operators. They believe in doing everything right. Everything should look right. No money is spared on making everything the best. Ken Grossman and Tim Cooper *care*. They are hands-on—accessible to all their employees and to their customers. They are big companies (Sierra Nevada produce more than a million barrels each year, while Coopers is around 0.7 million barrels). If "craft" has any meaning, then here are two shining examples, and yet the Independent Brewers Association in Australia excludes Coopers from membership for being too big.

I am not sure what the word truly is. It is not "craft." Perhaps it is "conscious." Or "conscientious." Perhaps it is "concerned."

2

Processes

He was a new recruit to the Brewing Research Foundation in leafy Nutfield on the North Downs of Surrey, England (Fig. 2.1). Scottish chap. We were sitting in the lunchtime sun, gazing out at the beautiful gardens.

"Would ye nae think that folks would have come up wi' something else to do wi' malt other than making whisky?"

I soon realized that he was in earnest. He truly did not know that the product his new employers were dedicated to studying consumed far more malt than any other process on the planet. Yet how many people know what malt actually is?

Indeed, when one asks people, "What is beer made from?" the inevitable reply is hops. In reality, hops are but a spice in beer. The bedrock of most beers is malt.

Time to take a stroll through the astonishing complexity that is the production of beer (Fig. 2.2).

Malting

Grapes are to wine what grain is to beer. The difference is that you just need to crush grapes with your feet to get out the sugar that can immediately be converted to alcohol by yeast. If you try crushing raw grain, you will just get sore feet. The levels of free sugars in grain are extremely low. Instead they are all linked together in enormous chains that are known as starch. Yeast cannot ferment starch. The starch needs to be broken down into its component sugars before the yeast is thrown in. Already you can see why beer is much more complicated to make than is wine.

The main cereal used for making beer globally is barley (Fig. 2.3), and we are referring to the grain that is in the ear of the barley (Fig. 2.4). The whole plant is important insofar as the roots and the straw are all essential in the growth of the plant, but it is the seeds that we are interested in for making beer. Except we only call them seeds when they are destined to be sown back

Fig 2.1. Lyttel Hall. These days a pricy apartment building but formerly home to the Brewing Research Foundation.

into Mother Earth to make more barley plants. For our purposes we call them grains, or corns, or kernels. Some people even call them berries.

There is a very good reason why barley historically is the cereal of choice for making beer. On the outside of every kernel there is a tough layer, called the husk or the hull, and this functions in the brewery as a filter bed when we are separating the liquid extract in the brewhouse. Other cereals don't have or don't retain this layer, and so they are rather more challenging to handle in the brewery. Even in those beers that are made primarily from other cereals—and here we might think particularly about wheat-based products—a proportion of malted barley is usually included in the mix to provide this husk.

If you get a handful of barley grains and pop them into your mouth, it's going to be something of a disappointment if not an outright shock. It's not like popping a few grapes into your buccal cavity: chances are that their sweetness fills you with good cheer. They might even be more palatable than the wine derived from them. But barley? Oh dear. Your mouth will start to dry out. Your teeth will know that they have been in a battle in trying to grind their way through the tough bullets of starch. Husk will lodge itself between your teeth. And as often as not you'll start coughing as the grainy powderiness will attack the back of your throat. Not a pleasant experience.

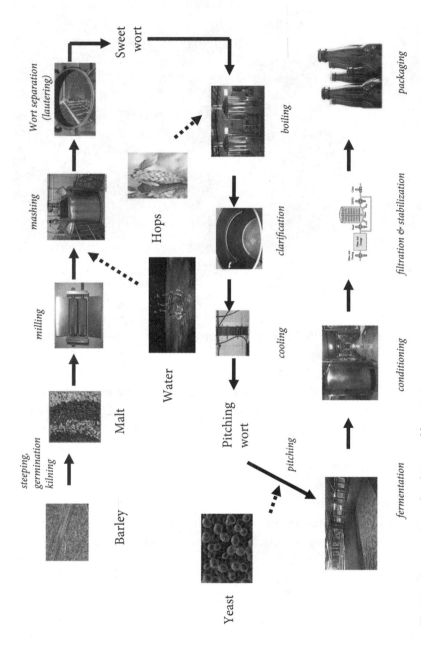

Fig 2.2. An overview of malting and brewing

Fig 2.3. A field of barley. Thanks to Muntons plc.

Hence the malting process (Fig. 2.5). This first happened 8,000 years ago by accident. Someone near to what nowadays is known as Iraq got the barley wet and it started to sprout (which is what a seed does when it gets enough moisture to get going). These ancients will have found that when they now bit into the grain, it was nice and soft and tasted like bean sprouts in a Chinese restaurant (not that they would have drawn that analogy back in the Fertile Crescent of 6000 BCE, but you get the drift). And then somehow they discovered that they could bake the crushed softened grains into a bread and when they dried it this bread developed even more pleasant flavors, going from the realm of raw vegetable to mellow caramel, and flavors like malted milk balls (again, the Whopper or Malteser had not yet been marketed, but that flavor was what they were now getting—and enjoying).

Through experimentation and perfection we have arrived at the traditional malting process that last saw any major change some 40 years ago with a change to the drying process that ensured that there was no risk of potentially carcinogenic substances called nitrosamines being developed. Apart from this, the three basic successive stages of steeping, germination, and

Fig 2.4. Holding ears of barley showing the grain. Thanks to Muntons plc.

kilning have remained largely unaltered for millennia. They have just been refined and made more efficient.

The barley itself is grown all over the world, with the main locations for producing malting barley being Australia, Canada, and several European countries. In the United States the number one producer of malt is Idaho, followed by Montana, North Dakota, Colorado, and Wyoming. Canada produces about twice as much as does the United States.

Not any old barley can be successfully converted into malt. Thus most of the barley on the planet is feed grain, which is used to fatten animals or, after removing the outer layers, to make pearled barley to thicken soup for the human being. Malting barley varieties are those that give high levels of fermentable material after they have been properly malted and brewed with. Countless scientific hours have been dedicated to researching why barleys differ in their malting quality, and we needn't dwell on these complexities here. I would just mention one

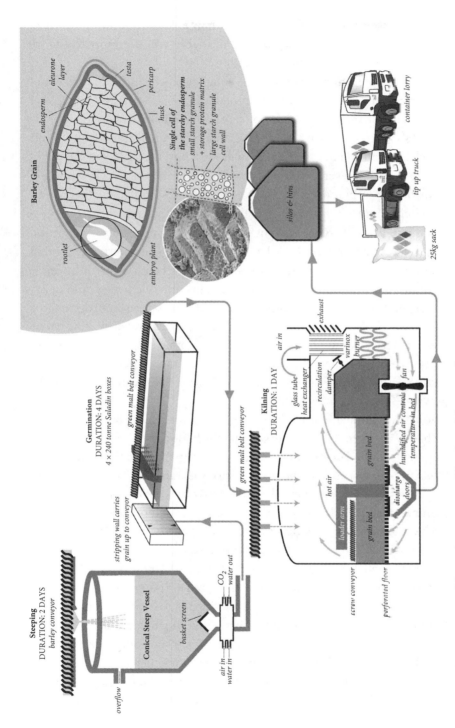

Fig 2.5. An overview of the malting process. Thanks to Muntons plc.

thing, which is hopefully quite logical, namely that the more starch in the malt, the more potential sugar is available and therefore the more potential there is for alcohol production.

Now there are a number of separate complex biochemicals in barley, and the two major ones are starch and protein. Imagine you have a couple of identical transparent boxes and two sets of colored balls: a red set and a blue set. Let's say red is protein and blue is starch. Say there is room for 100 balls in each box and you put 10 red balls in the first box, meaning you can put in 90 blue ones. However, if you put 20 red balls into the second box, there is room for only 80 blue ones. The more protein, the less starch. In other words, if Mother Nature fills up too much of the barley with protein, there is going to be less starch and therefore less potential for making beer. This is the reason why farmers and growers can't be throwing nitrogenous fertilizer around willy-nilly to increase the yield of malting barley, because that nitrogen is going to end up as protein (starch just contains three elements: carbon, hydrogen, and oxygen; protein contains these three—but also nitrogen and a little sulfur). This means that the yield per acre (or hectare) of malting barley is less than feed barley—but also less than many other crops. So growers have to get a cash sweetener (the "malting premium") to be persuaded to grow malting barley. As you can see, it is not only wine folks who obsess over the growing of their basic raw material.

So, back to the malthouse, which is logically most likely to be close to where the barley is grown. Having checked that the barley is alive and wholesome, the grain is steeped in water, which allows the kernels to kick into action (Fig. 2.6). After one to two days the steeped barley is transferred over to a germination tank, where over a period of three to six days the corns are allowed to germinate (Fig. 2.7). Modern-day malthouses have automated systems with mechanical means of making sure the grain is gently moved to allow good air flow, equalization of temperature and moisture, and prevention of matting of the rootlets that appear as a sign that the barley is growing (Fig. 2.8). Historically, this was all done in something called "floor malting," in which folks wandered up and down the floor on which the barley was spread, either thinning it out to allow it to cool down or piling it up to increase the temperature, all performed with rakes and shovels. There is a resurgence of interest in malting in this traditional way, with people going dewy-eyed when they speak of "floor malted Maris Otter" (the latter being a famous English variety of malting barley), but the truth is that this is all about romance and little to do with efficiency and malt quality.

Fig 2.6. Steep tanks. Thanks to Muntons plc.

This germination phase is a trade-off. The purpose of it is to develop the enzymes (biological catalysts) that will break down the complex substances that make the barley tough but which will also digest proteins to the appropriate extent and be ready to later break down the starch in the brewhouse. The consequence is the production of those rootlets and a shoot (maltsters call it an acrospire), which are an inevitable eventuality (after all, each of these little grains thinks it is going to grow into a fully-fledged barley plant). The trick is to allow the growth to take place just long enough to make the enzymes and allow them to make the grain more friable (millable).

Once the appropriate extent of growth has occurred, germination needs to be halted, and this is achieved in a third vessel, this one referred to as a kiln (Fig. 2.9). As the name would lead you to believe, this involves heat. Warm air is brought through the grain bed in a period typically of around a day to sweep out surplus moisture and bring the water content down to a very low percentage so that the grain is nice and stable. Too much heat tends to kill off many of the enzymes, so there is a gradual buildup of the temperature, these enzymes being more resilient to heat when the water level is lower.

Fig 2.7. A germination vessel. Thanks to Muntons plc.

Still other things happen on the kiln. There is the driving away of the raw vegetable flavors that develop in the germination, aromas such as bean sprout and cucumber. However, the most significant flavor changes occur in a chemical reaction named after a French chemist called Louis-Camille Maillard, who discovered it. Sugars that are produced in the germination phase meld with amino acids that are released from proteins to form interesting flavors and also color. The stronger the heating, the more flavor and color (Fig. 2.10). At first there is the development of those nice gentle flavors that we tend to call malty and which we associate with those malted chocolate candies and also with cornflakes, which don't taste of corn but rather of the malt extract that is sprayed onto them in production. You might buy yourself something like Horlicks or Ovaltine, powders that are mixed with hot milk to render

Fig 2.8. Sprouted barley. Thanks to Muntons plc.

a soothing bedtime drink. These are made from these so-called pale malts. Pale malts are the workhorse of the brewery, the soul of beer. They provide these gentle mellow flavors and relatively pale colors (hence the name) but they are also the main source of the enzymes needed by the brewer. For many very lightly colored lagers, for example, pilsners, these malts are extremely gently dried. For pale ale malts, through, a little more heating is used, giving that somewhat darker color and maltier flavor, but still the heating is not so high as to destroy enzymes unduly.

As we increase the heating, the flavors pass through into the caramelized region, and some more complex chemistry is going on that we needn't worry about. So now we have richer colors, in the redder, browner region and those tastes and aromas that make us think of caramel candies and toffee. These malts have had most of their enzymes destroyed so that they cannot be used alone for the making of beer, because the starch would not be broken down in the brewhouse and there would be no sugars for the yeast to work on to

Fig 2.9. A kiln. Thanks to Muntons plc.

make alcohol. Instead they are used as a proportion of the grist to give extra color and flavor. So maybe the brewer will use 85% of the pale malt, the work-horse, together with the balance of several of these specialty malts, which have names like caramel malt and crystal malt.

Some of these specialty malts are made with an extra process stage. The pale malt is transferred from the kiln to a roasting drum and the malt is heated intensely, just as one would roast coffee beans. Now we have still more complex chemistry going on as we produce heavily roasted products like chocolate malt and black malt, which have immense color and robust flavors like mocha, chocolate, and, yes, burned. They are used for the very dark beers, like porters and stouts, again as a minor percentage alongside the pale malt. Many stouts incorporate roasted barley, in which there isn't even a germination phase and the raw barley is simply burned. In my after-dinner speeches I tend to liken the flavor to that of smoking 20 cigarettes, stubbing them out in an ashtray, and licking the ashtray. Not that I have done that—but you get a sense of the harsh character developed. Having said that, I, like millions of others, delight in an Irish stout on the Emerald Isle. Just as we savor a dark-roasted coffee.

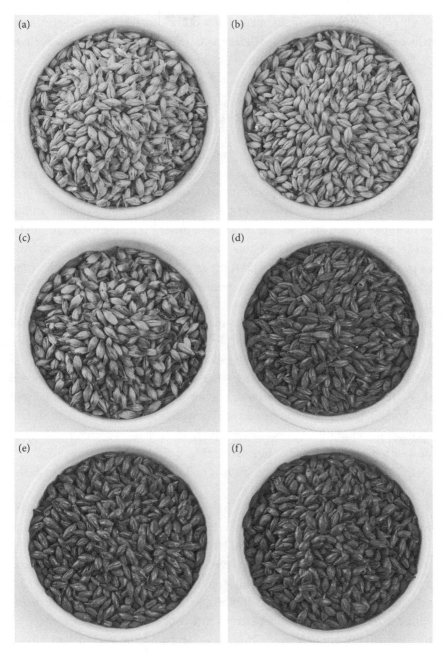

Fig 2.10 Types of malt: (a) pilsner, (b) pale ale, (c) crystal malt, (d) chocolate malt, (e) black malt, (f) roasted barley. Thanks to Muntons plc.

Of course there are malts made from other cereals: if it can germinate, you can malt it. So we have wheat malts (for wheat-based beers such as *Hefeweizen*), and malted oats, and malted rye, and so on.

Sweet Wort

The malt cannot be brewed with immediately after being stripped from the kiln. It needs to be stored for a few weeks; otherwise it will tend to perform badly in the brewery, giving poor rates of liquid flow.

The first thing that a brewer must do with the malt after storage is mill it to finer particles. This is so that the desired materials inside the grain can be easily extracted. However, that outer husk layer should be damaged as little as possible, because in as intact a form as possible it will go on to become the filter bed we spoke of earlier. So our mill needs to be set up in a compromise way, allowing us to preserve the husk while permitting the ensuing fragmentation of the inside of the grain (the endosperm) to the necessary small particles known as flour and grits.

For the majority of brewers this involves a roller mill (Fig. 2.11). In it a pair of rollers are set quite close together, and the grain is crushed between them. For a malt that has been nice and evenly germinated (we call it "well modified"), that setting can be reasonably wide, allowing the husk to simply split open without being unduly crumbled while that nice and crumbly endosperm is easily reduced to fine particles.

The sum total of all the materials that go into the mill is referred to as the grist—hence "grist to the mill."

Now the milled malt is mixed with warm water in a process known as mashing (Fig. 2.12). There may be three parts water to one part solids. The magic temperature is around 65°C (149°F). At this degree of heating the starch granules that have largely survived the malting process melt or paste in a process called gelatinization. (Think of making porridge in the kitchen: this selfsame process is happening.) This renders them far more amenable to enzymic attack. And this temperature is just cool enough for those enzymes to survive long enough to get the job done, in a time period typically of one hour. The bulk of the starch is broken down to produce a sugar called maltose (this name was given to this particular sugar because it is the one that is produced when malt is mashed. It actually comprises two glucoses attached together, but, again, we don't need to get too scientific. If you yearn

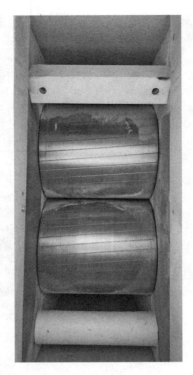

Fig 2.11. The inside of the UC Davis pilot brewery two roll mill. Thanks to Joe Williams.

for molecular structures, there's plenty of recommended reading at the end of the book.)

Now it is time to separate out the liquid (we call it sweet wort, because it is loaded with sugars) from the residual spent grains. This is most widely carried out in a lauter tun (Fig. 2.13), a very wide vessel at the base of which is a vast plate with holes or slits in it. The mash is pumped into it with the valves that allow liquid to pass out of the vessel in a closed position. The mash floats on the base and a bed is allowed to form on the plates. Then the valves are opened and the liquid allowed to start flowing. We can't go too quickly or the grains will tend to clog the holes. The wort is pumped over to the next vessel, which is called the kettle (or the copper, because historically it was made out of copper). The first worts contain a lot of sugar ("strong worts"). The problem is that pretty soon the liquid stops flowing and most of it remains wrapped around the spent grain particles. So we have to wash it through, and this is performed with sparge water, hot liquor that is sprayed over the grain

Fig 2.12. Inside a mash mixer. Thanks to Briggs of Burton.

Fig 2.13. Inside a lauter tun. Thanks to Briggs of Burton.

bed and which percolates through, washing out the sugars into the kettle. Once we have collected the necessary amount of wort in the kettle, it is time to empty the lauter tun for the next batch. Those spent grains will get contaminated very quickly, so we need to get them out of the brewery and to the nearest dairy herd: cows love them.

Water

A word about water. After all, it is the major component of all beers (except those of ridiculously high alcohol content). Winemakers rejoice in the term terroir, but if anyone might legitimately use this expression it would be the brewer. The water composition is very much dependent on the minerals that are present in the ground from which the water is sourced. Thus we speak of hardness, which refers to the level of calcium (and to a lesser extent magnesium) that is present. If the water comes from a region rich in, say, gypsum, then it contains a lot of calcium sulfate and is referred to as permanently hard water. If the water is from a terrain that is rich in carbonates and bicarbonates such as chalk, then it will be rich is calcium bicarbonate and we have temporary hardness, so called because it can be removed by boiling. There will be many other minerals besides. These various materials can profoundly impact the properties of the water and thus of the brew and the beer. At one extreme we have the water from Pilsen in the Czech Republic, which is amazingly soft, with very low levels of dissolved salts. At the other extreme is the water from Burton-on-Trent in England, which is astoundingly hard. To try to make the same beer using these two extremes of water would be fraught with difficulty. So what brewers do if they are trying to make the same beer in breweries across the world is adjust the water composition. You can remove salts from water using certain filtration techniques. You can add salts to water from big bags to boost levels of things like calcium. In Germany they even have a word for this, Burtonization. The important thing is, you have to regulate this salt content if you want the process to proceed the same way every time and if you want the beer to taste and smell as you expect it to.

Brewers seem to be falling over themselves to see who can use the least water. The best performers are in parched Queensland, Australia (Fig. 2.14), where they are using not much more than two barrels of water for every barrel of beer—remember that you need water for things like heating (it is converted to steam to pass through jackets on vessels) and cooling, as well

Fig 2.14. The Yatala Brewery. Thanks to Tom Robinson, Carlton United
Breweries.

as to clean. A reasonable global achievable target is three barrels of water for
every barrel of beer. I would just say that brewers are a wee bit naughty in
focusing only on brewery water usage; for some it has become almost a sales
pitch. The reality is that vastly more water is used in growing crops like barley
and hops and, of course, in the malting process.

Wort Boiling

To the kettle, then, and the most energy-demanding stage in the brewery
(Fig. 2.15). This is how beer saved the world, because it is in the boil where
microorganisms are killed. Through the ages people got sick drinking contaminated water, but those who drank beer did not, because of this boiling
stage in the process. Other things happen in the boil, including the driving
off of undesirable aromas (what you can smell when you are near a brewery
is the boil; it's a pleasing enough whiff but not one that we would likely
want in our pint). There is also the evaporation of water, something that
is particularly important if we want to make stronger beers such as barley
wines. Historically these worts are boiled for a long time to drive off water,

Fig 2.15. A brew kettle. Thanks to Briggs of Burton.

concentrating the sugars so that later the yeast will make more alcohol. In the boil there are also some flavor changes taking place, giving fuller character to these types of beer. A further occurrence in the boil is the precipitation of proteins (analogous to how egg white curdles if you boil it) to make something called hot break.

Historically, though, perhaps the most important role for boiling was (and still is for most brewers) the extraction of bitterness from hops. Let's just dwell on hops for a while.

Hops

Hops, *Humulus lupulus* (a close relative to *Cannabis sativa*), are dioecious plants, that is, there are males (Fig. 2.16) and there are females (Fig. 2.17). It is the females that interest the brewer because they develop the flowering entities called cones. Within the cones is the so-called lupulin, sticky yellow glands that contain the resins (which will give us the bitterness) and the oil that provides the characteristic hoppy aroma (Fig. 2.18).

Fig 2.16. Male hop plants. Thanks to Jay Prahl, Hopsteiner.

The number one growth location for hops globally is the United States, with Germany a close second. But hops grow well in the latitudes between 38 and 53 degrees north and south of the equator. So there are famed hop-growing regions in places like Tasmania, New Zealand, and, yes, England. In the United States by far and away the largest amounts of hops are grown in Yakima in Washington, followed a long way behind by the Willamette Valley and Grant's Pass in Oregon, and in Idaho.

Hops grow on trellises to a height of up to 16 feet and present a not inconsiderable challenge in cultivating (Fig. 2.19). And there are many

Fig 2.17. Hop cones. Thanks to Jay Prahl, Hopsteiner.

Fig 2.18. Cross section through hop cones. Thanks to Jay Prahl, Hopsteiner.

Fig 2.19. Preparing the hop yard in the spring. Thanks to Jay Prahl, Hopsteiner.

hop varieties. We can divide them into bittering, aroma (or noble), and dual-purpose hops.

Bittering hops contain a lot of resin and therefore have huge bittering potential. The aroma hops tend to be low in resin, but their oil is far more pleasing than that from the bitter hops. Dual-purpose hops have both respectable amounts of resin and pleasing aromas.

There is a burgeoning number of hop varieties these days, and thankfully an improved vocabulary around the diverse aromas that they are capable of delivering (Fig. 2.20). But it takes great skill from the brewer to make the most of these hops.

For bittering proposes the hops are traditionally added at the start of the boil and, with a traditional cooking time of around an hour, there is ample transformation of the main resin, alpha acid, into the chemically rearranged, more soluble, and more bitter iso-alpha acids that are the major bittering entities in beer. The problem is that this conversion is not totally efficient, and furthermore there is a sizable loss of the iso-alpha acids downstream of the kettle: they tend to stick onto surfaces and onto

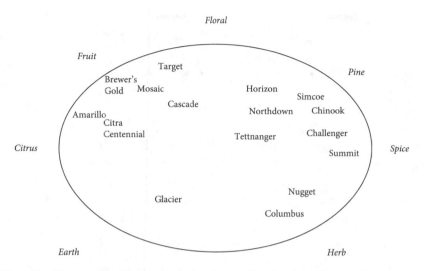

Fig 2.20. The aroma of hops. An approximate distribution of hop varieties according to their primary aroma notes.

particles, not least yeast. Accordingly, the amount of bitterness measurable in beer is really rather less than the potential bitterness that is added at the start of the boil. That bitterness in beer is referred to as International Bittering Units (IBU) and is measured by seeing how much ultraviolet light is absorbed by an extract from the beer. For whole-cone hops put into the start of a boil, the chances are that only about 25% of the potential bitterness is realized as IBU in that beer. However, that is a price that some brewers, notably Ken Grossman at Sierra Nevada, are prepared to pay to use a material that they feel should be as close to nature as possible. Indeed, I am sure that if Ken had his druthers he would always use newly picked hops, analogous to how the winemaker makes all of his or her product at the time of the harvest, the crush. However that is not realistic for brewing, which is a year-round activity, so the hops are dried to render them more stable. (Some beers are produced at the time of the hop harvest, and these are labeled fresh hop or wet hop.) Even that drying will take away some of the aroma from the cones, but it is a price that needs to be paid. Those hops must be packed tight to exclude oxygen that will make them turn cheesy, and also they must be stored cold. One of the most delightful rooms in a brewery such as the astonishingly beautiful Sierra Nevada facilities in Chico, California, and Mills River, North Carolina, is the hops store. What a sublime assault on your nostrils.

Fig 2.21. Hop pellets. Thanks to Jay Prahl, Hopsteiner.

Most brewers, though, seek to recover as much bitterness as possible, so many use hops after they have been dried, hammer-milled, and extruded into pellets (Fig. 2.21) that disintegrate in the boil and give a greater opportunity for the resins to be extracted and transformed into iso-alpha acids (so-called isomerization).

For some brewers, though, even that is not good enough. They use something called post-fermentation bittering. It was shown a good many years ago that you can extract hop powder with liquid carbon dioxide and make a liquid that can be chemically transformed in a factory to produce so-called pre-isomerized extracts (Fig. 2.22). So now, rather than extracting bitterness in the boil, you can add the bitterness to the end product and circumvent all the process losses. Fosters in Australia has long since championed this approach. The Miller Brewing Company goes one step further, adding hydrogen atoms to the iso-alpha acids to make so-called reduced iso-alpha acids. Doing this makes a form of bitterness that does not break down in the presence of light to give skunky flavors. Regular iso-alpha acids succumb to light very rapidly, which is why beer is most sensibly packaged in a brown glass bottle, which color does not allow the damaging light to get into the beer anywhere near as rapidly. Green glass and the clear glass favored by Miller don't hold back the light. Mind you, a can is best.

Fig 2.22. Hop extracts. Thanks to Jay Prahl, Hopsteiner.

Fig 2.23. Rubbing and sniffing. Thanks to Jay Prahl, Hopsteiner.

What of the aroma from the oils? These oils are very volatile, which means they evaporate very readily. One of the most abiding images of brewers is their rubbing hops between their hands and smelling their palms (Fig. 2.23). What is happening is that the lupulin glands are being disrupted and the warmth of the hands is sufficient to drive the oil molecules into the sniffer's nose. Small wonder, then, that if you put hops into the start of a boil, all of that aroma is blown away up the stack. Accordingly, there are two general approaches to introducing hop aroma into beer. The first of these originated in the production of Czech Pilsners. Here most of the hops are added at the start of the boil to get out the bitterness. But a smaller proportion is introduced late in the boil (hence "late hopping") to allow some of the aroma to survive. In turn some of the molecules are transformed by yeast so that in the finished beer we have an often subtle hop nose. The other approach we can trace back to the production of English cask-conditioned ales and the addition of a handful of whole hops into the finished beer, so called "dry hopping." This allows for a greater proportion of the aroma to survive. Now in North American craft brewing circles, that handful has become what can seem like a truckload, leading to immense hoppy noses in beers like American IPAs. In Sierra Nevada they have torpedoes, for which more than one of their beers is named. These are torpedo-shaped vessels packed with hops and through which the cold finished beer is recirculated to thoroughly extract hop aroma.

Clarifying and Cooling Wort

Back to the brewhouse and the stage that follows the boil. This will vary, depending on the mode of hopping. If whole-cone hops are used, then we have a hop back (or hop jack) in which the wort is filtered through a bed of residual hops. If, however, the brewer has used pellets or a liquid extract of hops, then the clarification is accomplished using a whirlpool (often called a hot wort receiver, Fig. 2.24).

To understand this piece of kit we need to transport ourselves back to the kitchen of Albert Einstein. The venerable physicist was making himself a mug of tea in the days before tea bags. He would plop a spoonful of tea leaves into the mug, add boiling water, and stir. Being a scientist who liked to investigate things, he watched the tea leaves as they swirled around and then gathered in a pile in the middle of the base of the mug, and he said, "Wow, what a great idea for the brewing industry." Actually he probably did not say

Fig 2.24. A whirlpool. Thanks to Briggs of Burton.

any such thing, but he did write up his observation for publication, and his work was noted by one H. Ranulph Hudston, who invented the whirlpool for the Molson Brewing Company in Montreal, Canada, in 1960. Basically the newly boiled wort is sent into the whirlpool at an angle, and it swirls around gently, such that the material precipitated out in the boil settles in the middle of the base of the vessel. The "bright" wort can be pumped away through pipes and the residue collected and mixed with the spent grains before they head off to the farm.

Now the wort needs to be cooled down because yeast really does not like to be flung into wort at temperatures approaching 100°C (212°F). Historically this chilling was achieved using a so-called coolship (Fig. 2.25), in which the boiled wort is pumped to a shallow vessel located at the top of the brewery in which the wort is allowed to cool naturally. The steam coming off the brew escapes through slats in the roof, but of course debris of various types (including from the posteriors of passing birds) can head in the opposite direction, so this is not necessarily the most hygienic of processes. It survives to this day, though, in some breweries, notably the makers of sour beers such as Lambic.

Fig 2.25. A coolship. Thanks to Russian River Brewing Company. This is the coolship in the company's newer location in Windsor, California.

Fig 2.26. The two paraflow heat exchangers in the pilot brewery at UC Davis in the foreground. The first cools using cold water; the second cools using propylene glycol. Thanks to Joe Williams.

Nowadays it is much more common to have a paraflow heat exchanger (Fig. 2.26) which is akin to a car radiator. There is a multitude of very thin plates, on one side of which hot wort flows and on the other there is a flow of cooling liquid heading in the opposite direction. Heat passes across the plate so that the wort cools down and the cooling liquid heats up. The wort makes its way to the fermenter while the warmed coolant has the energy collected from it and is then reused. These cooling liquids are either cold water, which is able to take the temperature down to the range needed to ferment ales, and either propylene glycol or ammonia, which lowers the temperature to the significantly lower range historically used for lagers.

Fermentation

And so we come to fermentation. We need to add to the cooled wort two main things: yeast and oxygen. The yeast needs the oxygen to make the membranes from which it is constructed. The yeast itself is a unicellular fungus (Fig. 2.27). There are two types: ale yeast and lager yeast. You can go hunting for the former all over the planet and you will find it in lots of places, especially those rich in sugar, such as fruits. However you simply will not find lager yeast at all. It is an amazing organism that emerged probably within the last couple of hundred years by two yeasts coupling in a laboratory or a brewery. I like to say that somebody dimmed the lights and played soft music to encourage this copulation but, whatever the circumstances, these two strains, *Saccharomyces cerevisiae* (ale yeast) and a wine yeast called *Saccharomyces eubayanus*, merged to form lager yeast, a more complicated organism that taxonomists these days call *Saccharomyces pastorianus*. And so there are relatively few lager yeasts and vastly more ale strains.

The fundamental difference between an ale and a lager comes down to which of these yeasts is used to carry out the fermentation. Ales are made with ale yeasts, in fermentations that are usually at higher temperature than for lagers because ale yeasts are more thermo-tolerant. Lagers are beers made with lager yeast, usually under cooler, more prolonged processing conditions.

Traditional fermenters used to produce ales in shallow, "open" vessels (Fig. 2.28). The ale yeast rises and can be skimmed off and used to inoculate the next batch in a separate fermenter.

These days many beers, both ales and lagers, are produced in sealed cylindroconical vessels (Fig. 2.29).

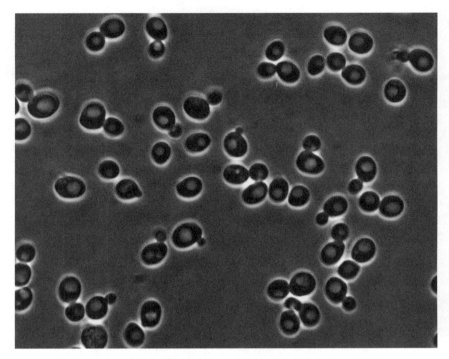

Fig 2.27. Brewing yeast. Thanks to Kevin Verstrepen, KU Leuven.

Fig 2.28. An open square fermenter. Thanks to Simon Yates, Marston's.

Fig 2.29. Cylindroconical fermenters. Thanks to Jamil Zainasheff of Heretic Brewing Company. If you look closely under the word Heretic, this fermenter is called Bamforth. "Ah shucks," I said, "why did you call it that?" Jamil replied, "Because you're the biggest heretic we know."

What happens in a fermenter is that the yeast, boosted by the oxygen enabling it to reproduce, takes the sugar from the wort and in a rather complex series of metabolic reactions converts it into ethanol (alcohol) and carbon dioxide. It also produces more yeast—about three times more cells at the end of fermentation than at the beginning. The yeast also modulates the flavor, removing some undesirable substances from the wort, but making a range of new ones that have a range of flavors such as fruity (the esters) and vegetable (the sulfur-containing substances).

Fig 2.30. Marmite. Photo by the author.

Fig 2.31. Vegemite. Thanks to Peter Aldred, Federation University, Australia.

The brewer monitors the progress of the fermentation by measuring the strength of the wort as specific gravity. (The specific gravity is the weight of a given volume of wort as compared to the weight of the same volume of water. One milliliter of water weighs one gram, but if sugars are dissolved in the water, this raises the specific gravity. So the wort might be said to have a specific gravity of 1.04, for instance.) As the sugars are removed and alcohol is produced, the specific gravity goes down (this is on account of the removal of sugars, but also because of the production of ethanol, which has a very low specific gravity of 0.79). Once the specific gravity has tailed off, we know that the yeast is done making alcohol.

However, that is not usually the end of the process in the fermenter. Yeast also makes a substance called diacetyl. This stinks of popcorn or butterscotch and is a definite no-no for most beers (though there is a famous Czech beer in which it is supposed to be present and you can certainly detect it in many a chardonnay). The good news is that yeast will mop up this diacetyl and take it back into the cells to eliminate it. But that takes time. So good brewers will check for this diacetyl, either instrumentally or using their nose, and once it is gone, they are ready to ship the liquid to the next stage, once they have collected the yeast, which often goes to pig food; or to pitch the next fermentation; or to Marmite (Fig. 2.30) in the United Kingdom or Vegemite (Fig. 2.31) in Australia.

Cold Treatments and Packaging

And that next stage is called cold conditioning. Very often the yeast is removed using a centrifuge and then the beer is taken to the lowest temperature possible without freezing it. In practice that is often around −1°C, at which temperature beer does not freeze because alcohol lowers the freezing point. At this temperature some proteins and tannins start to appear out of solution and they combine and settle out. This is good, because if they are removed at this stage, they are not going to appear as haze in the finished beer. Some brewers will exaggerate the removal of haze-forming materials by using agents that remove either tannins (using something called PVPP) or proteins (using either tannic acid, or silica gels, or enzymes).

After a few days the beer may be filtered (though not always) and sent to packaging into either bottles or cans or larger-pack containers such as kegs (Fig. 2.32). Packaging is very much the most expensive stage in the operation,

Fig 2.32. Packaging equipment: (a) a bottling line (thanks to Terence Sullivan, Sierra Nevada), (b) a canning line (thanks to Simon Yates, Marston's), (c) a kegging line (thanks to Simon Yates, Marston's).

and truly beer quality can be jeopardized if it is not done properly. For small containers the beer is filled into cans or bottles using mighty rotary fillers that draw a vacuum in the container then fill in the right amount of beer and finally apply either a lid or a crown cork. Canning rates can exceed 2,000 cans per minute and bottling lines 1,200 bottles per minute. Each container will have the right amount of beer, the right carbonation level, and as little oxygen as possible (oxygen being the primary cause of staling). Highly skilled stuff—and very demanding. Then the packages will be coded and put into secondary packaging, such as cases, before heading to the warehouse and thence onto transportation.

And that's all there is to it.

3

Sorts

I talk to many Rotary Clubs. It means that I get a lot of practice singing (inter alia) "God Bless America," always struck by the mention of foam. To the point though: I am well aware of the most frequently asked questions from a beer layperson. High up on the list is, "What is the difference between a beer and a lager?" Straight away I have to point out to the inquisitor that beer is the overriding term and that within it we can divide the products up into ales, lagers, and other. I know that the questioner has an image of lager as being a pale yellow liquid and a mindset that the terms "lager" and "Pilsner" are interchangeable. I disabuse them of this quaint notion and point out that there is a rich diversity of lagers, including very black ones. The unifying attribute that makes them lagers is that they are fermented using lager yeast, whereas the thing that makes an ale an ale is that fermentation is effected using an ale yeast. And the others? Well, those are the beers made using other types of organism above and beyond an ale or a lager strain alone. Such yeasts may be part of the mix, but there is a diversity of other microbes involved in the process, both fungi and bacteria. A prime example would be Lambic.

The word "lager" actually means "to store" and is founded on the original way of making such products, which involved prolonged holding in cold cellars. Although there are still some brewers who insist that this is the only decent way to make a great lager, there are plenty of others who beg to differ.

In addition, when it comes to the word "ale," its meaning is very different now from what it was in times gone by. Prior to the introduction of hops into the British Isles, the products in my native land were referred to as ales, and were of necessity strong in order that the alcohol could inhibit the growth of those microorganisms that would seek to spoil the brew. Those products that started to include these newfangled hops from the mainland of Europe ("a wicked and pernicious weed" was the view in the early 16th century) did not need to be so alcoholic because the hops contain antibacterial agents. (St. Hildegard von Bingen [Fig. 3.1] drew attention to that.) In those days, these hopped products were called "beer," as opposed to the unhopped "ale." These

Fig. 3.1. St Hildegard von Bingen. Courtesy of Jay Brooks.

days, of course (thank goodness), the vast majority of beers, whether ales or lagers, are hopped.

The successful brewing companies are characterized by strong new product development programs, from which have emerged some remarkable beers that do not fall easily into any recognized classification. In addition, the burgeoning craft beer sector seems to believe that nothing is sacrosanct or unworthy of exploration. Where, for instance, would you pigeonhole a stout containing oysters or chocolate and, perhaps most ludicrously of all, Rocky Mountain oysters (aka bull gonads) or the fluid produced in said organs?

Even more fundamentally, beers that may fall into an obvious genre in one market may be slotted into an entirely different category elsewhere: for instance, a beer that may be described as a "bitter" in Australia would to an Englishman be perceived as having the characteristics of a "lager." I recall a beer brand in England called Long Life that started its "career" as an ale but, when the English brewers finally discovered what others had known about for a long while, namely lager, suddenly Long Life was reclassified into the sexy new (*sic*) genre. Indeed, it was also platformed as a beer designed for the can (i.e., the package type and not the room in which to drink it).

I know of more than the occasional brewer who only has one yeast strain working in the brewery but produces a range of brands that are marketed as ales or as lagers.

For the purist it is of utmost importance to label a product accurately. However, in these days that might even be described as "cross-dressing" days for beer, does it really matter? The important thing is to make products that sell and not necessarily those that rigidly adhere to the admittedly admirable set of guidelines laid down by the likes of the Brewers Association (go to www.brewersassociation.org/educational-publications/beer-styles/).   Later in this chapter you will read about a strong example of this when I talk about *Hefeweizen*.

Before we stride through the diversity of ales and lagers, we should ponder for a moment the impact that the application of science has had on new product development in recent decades. Whether they are ales or lagers is really rather secondary to the technology used to produce them or the story line told about them.

Consider, then, the beer introduced by one famed North American brewing company in 1993. Surely this was a classic example of technology push over market pull. You can imagine it, can't you?

Technologist: "Hey, I can take all the color out of the beer without changing the flavor or foam."
Marketer: "Great, we'll call it Clear Beer."

They did it—and the product was pulled back from the trade after just a few weeks of lackluster sales. Folks did not get it (in any sense of the term) any more they got the clear version of a famous cola developed at the same time. You pour out a liquid that looks like soda water but has a head on it, well that head is perceived as a "scum." If it has a color that is in the beer range then that is not a scum, it is a "foam." I rather suspect that the company thought that it would appeal to women. If that was their rationale then they fell afoul of a fundamental truth: *never* develop a product on a gender-targeting basis, whether it is style, volume . . . anything.

On the other hand, that same company did prove itself capable of glorious successes. Of particular note was the first commercially successful light beer. In fairness, it was not the technical people who came up with the idea, but the marketing folks who realized how to make the concept take off. The scientist behind the idea was Dr. Joe Owades. The Rheingold Company of Brooklyn product developed by Owades was Gablinger's. Soon after came Meister Brau from the Peter Hand Brewing Company in Chicago. The problem was not the beer, but rather the marketing, which was tantamount to "Hey, you're fat; you should drink this." Not cool. This brand found its way to the aforementioned major company, which filmed the rebranded beer in ads featuring a famous footballer . . . and now it was okay to drink these products brewed to contain fewer calories. Mind you, they could not spell "light."

Ice beer, dry beer, non- or low-alcohol beer: these are all concepts developed in the past half century with varying degrees of success. We will visit them later in this chapter, but as we pass to a consideration of the classic beer styles, let me remind you that there is an extremely long pedigree of developing beverages featuring beer that are founded on blends with other drinks. For example, ponder braggot, a hybrid of beer and mead.

Top Fermentation Beers

English Pale Ale

It is pale ale when in a bottle or "bitter" when on tap (draft—or "draught," as it is spelled in the United Kingdom). In the United Kingdom, tax is raised on beer in proportion to alcohol content (ABV), so there is a tendency toward the lower end of the ABV range for these (and most other) UK beers. Frequently you will find they are in the 3.5% to 4.0% ABV span, but usually are so beautifully balanced in flavor that they are deliciously drinkable, and so it is just as well that the low alcohol also means fewer calories, alcohol being the main source of calories in any beer. On tap they may be cask-conditioned (see chapter 5) and drawn using a beer engine relying on siphoning; or kegged, in which they will be pushed out of the container using carbon dioxide pressure. The former are never pasteurized and are of lower carbonation, whereas the latter are frequently pasteurized, usually filtered, and contain more fizz.

In the 1960s and 1970s there was a big shift toward the keg version, which demands less careful handling of the beer in the cellar. In defense of the cask style, the Campaign for Real Ale (CAMRA, see also chapter 1) came into being in my native northwest of England in March 1971 with the avowed aim to champion cask-conditioned products. Very laudable, for a good cask ale is indeed something well worth celebrating. However, it is unfortunate that CAMRA's stance (relaxed a touch in recent years) is that by definition anything other than cask ale is not worth drinking, especially lager. A trawl through the world's beers shows just how parochial such a mindset is.

English-style pale ales tend to be produced with a grist featuring a pale malt as the core component, usually boosted with some caramelized malts to add color and a depth of flavor. They will not be excessively bitter and will generally tend toward being more "malt forward" than hop forward in taste and aroma.

Some people subdivide pale ales into ordinary, special, and extra special—witness in the last instance, for example, Fuller's ESB from the celebrated Fuller, Smith & Turner brewery in Chiswick, founded 1845. Fuller's ESB (5.5% ABV) was CAMRA's first champion beer. (As I write, this company has been acquired by Asahi Breweries of Japan. We consider mergers and acquisitions in chapter 9.)

American Pale Ale

These beers are founded on a tradition of the English pale ale, but are generally "bigger" in alcohol, bitterness, and hop aroma. They may typically be in the 5%–6% ABV range, with bitterness level perhaps 1.5 to 2 times higher than their antecedents from the British Isles.

Another style to emerge through the "craft" beer evolution in the United States is *amber ale*, an ill-defined genre that is probably most generally expected to have a color closer to an English brown ale (see later) than a pale ale, a color delivered through the use of caramelized malts that also bring a rich maltiness to the brews. This maltiness is balanced with quite robust hop character.

At the other end of the color spectrum are the *blonde ales*, with a very pale yellow color afforded by gently dried malt, perhaps Pilsner malts. They possess the more bready notes associated with pale malts, this accompanied by a relatively modest hop character. However here, just as in pretty much any beer style, the latter-day brewer does not feel the need to be encumbered by any strict rules. Brewers might well call a beer "blonde" simply based on its color and feel perfectly entitled to do whatever they want in search of a point of difference and a new experience for the consumer. Thus they may barrel-age the product, employing "funky" microflora.

India Pale Ale

The history of India pale ales (IPAs) is generally presented in a somewhat ro-mantic fashion, which generally goes along the lines of their being designed to survive the passage by sea to outer regions of the British Empire, notably the Jewel in the Crown, India. Accordingly, they were rendered more bitter and more alcoholic, these properties affording greater resistance to spoilage. The Raj recognized the merit of bitter substances to counter diarrhea—hence gin and tonic, with its combination of alcohol and quinine.

The reality is that many beer styles were exported from Britain to the col-onies, including porter, though the IPAs from companies such as Hodgson, Allsopp, and Bass certainly enjoyed huge success, being rather more re-freshing than porters, for example.

Actually, alcohol and bitter acids did not prevent yeasts from growing. Indeed such beers were "conditioned" using *Brettanomyces* and will have been quite bold in barnyard flavors. Traditionally the IPAs would not have had particularly pronounced hop-derived aromas. IPAs in the modern North American tradition, to the contrary, have big hop noses as well as big bitterness and big alcohol. They might be best described is turbocharged pale ales. Indeed, there are now *double IPAs* and *triple IPAs*. There has even been a latter-day division of American IPAs into *West Coast* and *East Coast* genres, the former being relatively bright (and more bitter) and the latter having im-mense turbidity, to the extent of appearing as if they are merely highly hoppy thick suspensions of yeast or other insoluble materials. The East Coast style is not especially bitter but gigantic in fruity noses. They are sometimes lik-ened in aroma terms to unfiltered juices; indeed they are often referred to as *juicy IPAs*.

Another latter-day iteration has been *black IPA*, in which the grist includes some heavily roasted grains in the malt bill. Such products combine the rich coffee-like notes of a dark stout with the intense hop aroma of a double or triple IPA. They certainly meet my rule that a beer should be balanced in flavor. That balance might be in a relatively subtly flavored North American lager. Equally, it can be in a "big" beer such as a black IPA.

The most recent IPA iteration is so-called *brut IPA*. These often involve the addition of an enzyme called glucoamylase, derived from the fungus Aspergillus that is used in the production of sake. Adding this enzyme to the fermenter means that all of the starch is rendered fermentable and therefore very little carbohydrate survives into the beer (cf. the context of the word "brut" in terms of very dry champagnes). This increased fermentability also means more alcohol, of course. This enzyme is also used in the production of certain other stronger beers, such as imperial stouts.

Mild

This style of English ale is in decline (from a position in 1930s Britain when it accounted for 75% of the beer consumed). Mild tends to be sweeter and usually (but not always) darker than pale ale, with color and flavor from the use of caramel malts. Milds tend to have relatively low alcohol contents, often in the 3%–3.5% ABV range. Brown ales are essentially bottled milds. The original use of the term "mild" was to indicate fresh and relatively young beer, as opposed to beer that was at least in part aged (*stock ale*).

Mild ales were really in the ascendancy in the days of metal foundries in the Midlands of England, and a notable example of the genre is Banks's Mild from Wolverhampton. (Pronunciation thereabouts: "Bonksies"). This specific brew was the first pint I ever had when I arrived at university in 1970. I paired it with a bag of crisps (chips to Americans), so I guess a not particularly well-balanced lunch.

The low alcohol and pleasant sweet flavor of mild make it a highly drinkable product—and in these days when many brewers are looking for "sessionable" products, it might be high time to resurrect this genre. Hopefully it will be treated with more respect than was once the case in my homeland. I recall with horror my first day as a young barman at The Brown Cow in Hapsford being shown what to do with the "slops." This was the beer that overflowed when one pulled a pint, collecting in the tray underneath the spout. Through the evening you would keep emptying this beer into a bucket. Once the bar had closed you would go down into the cellar with the buckets and pour them back into the barrel of mild—never into a lighter colored beer! This practice was strictly forbidden by the company—but the manager was a cunning so-and-so and presumably on to a good thing for his own back pocket as he made the beer stretch further, pocketing the extra.

Scotch Ale

More malt and less hop forward than English ales and customarily sweeter, Scotch ales' strength has historically been designated in terms of the shilling (/-), the old British currency. (One shilling equates to five modern pence. There used to be twelve pence to the shilling and twenty shillings to the pound in the pre-decimal days pre-1971.) The stronger the beer, the higher the tax payment to the exchequer in terms of shillings paid. Thus beers could be ranked as low as 60/- (a *light* ale, not to be confused with the term "light" in terms of lower carbohydrate beers) and then at 10s. increments through a range of *export* ales, up to the 100/- to 160/-, *wee heavies*. The Original Extract equivalencies of these beers is 60/- (up to 8.75°P), through 100/- (18°P), up to 160/- (as high as 30°P).

Irish Red Ale

This is a term more to do with latter-day marketing of specific brands in the United States than any distinct historical provenance in Ireland, although the antecedent of the Coors brand Killian's Irish Red was a ruby ale from county Wexford. Such beers have a reddish hue and a substantial caramel note because of the use of caramel malts in the grist. There may also be a hint of burned character from a modest proportion of more strongly roasted grains.

Barley Wine

Very strong ales, typically close to 10% ABV, they are historically to be found in small bottles ("nips," containing about 190 mL of beer), though it seems to be increasingly common in the United States for them to be packaged in much larger bottles—hopefully for people to share. The strongest "first runnings" of wort from the mash are used, further concentrated by lengthy boiling and the addition of extra sugars in the kettle. All of this leads to the development of high levels of alcohol, flavor, and color. When yeast ferments such strong worts, it produces very high levels of fruit-flavored molecules called esters. This is one of the few beer styles that actually benefits from aging, and at least one brewer I know rejoices in presenting to the consumer the experience of "vertical tasting" of their celebrated barley wine, illustrating how the nuances of flavor change with time. *Wheat wines* are the equivalent made from a grist featuring at least 50% malted wheat.

Old Ale

A style of beer prominent in 18th- and 19th-century England and which was originally of similar strength to the barley wines was *old ale* (also known as *stock ale*). As the name would imply, this was a beer that was stored in wooden barrels for prolonged periods, inheriting some flavor characteristics from the wood, as well as doubtless acquiring some notes from adventitious microorganisms. The old ale was mixed with younger beer in the pub to deliver the blend desired by a given customer.

Naturally Conditioned Bottled Beer

A few brewers (notable examples are Coopers in Australia and Sierra Nevada in the United States) deliver many of their bottled and canned beers into specification for carbon dioxide via "natural conditioning." This is akin to what happens in the production of cask ales (see chapter 5), without the addition of hops or finings in this case. The beer is packaged with a significant number of yeast cells remaining, together with a small amount of sugar, the addition rate being calculated in relation to how much carbon dioxide will be made from it when the yeast metabolizes it. One of the most famous beers produced through natural conditioning is Worthington White Shield from Burton-on-Trent in England. In this beer the yeast is readily apparent at the bottom of the bottle, but such is not the case for most other beers of this type. The yeast employed may not be the same strain that is used for the primary fermentation of the beer.

Beers of pretty much any style can be conditioned in the package, so it is of itself not a separate variety of beer.

Porter

Porter was first brewed in the first quarter of the 18th century in England. Taxation of beer in those days was based on the quality of the raw materials employed. Malts dried by fires from untaxed wood attracted less duty than the "cleaner" malts kilned using coke (the government was concerned about the unhealthy impact of burning coal in urban surroundings). The cheaper malts, notwithstanding their coarse, woody flavors, could be used to produce cheaper beers affordable by the poorly paid masses, including the porters in the markets of London. The beer brewed from the paler malts were twice as expensive—*tuppence* (two pence, there being twelve pennies in a shilling and twenty shillings in a pound). Plenty of hops were also employed, affording durability as well as pungency.

Stout

The London-based ale brewer Michael Combrune wrote a book called *The Theory and Practice of Brewing* in 1804. In the 1750s he had been the first to employ a thermometer in a brewery, and he was able to see how he got greater yields of fermentable wort from malts that had not been heated strongly. (We now know all about the impact of heat in destroying the enzymes that break down starch; see chapter 2.) And so rather than using only these coarsely dried malts, brewers shifted to employing a mixture of pale malts and black malts produced in roasting cylinders patented by Daniel Wheeler, a sugar roaster from London's Drury Lane, in 1817—hence the synonym for black malt being "Patent Malt"). Rather than coarse, woody notes in the darker brown ales and porters, these beers made from pale and patent malts possess more roasted coffee character as well as a black as opposed to brown coloration. The "extra" dark (and stronger) variants were "extra stout porters," with the "extra" and "porter" to be dropped such that we now had *stout*. I would contend that stout is the only beer whose flavor is benefited by the presence of nitrogen gas, which was pioneered by Guinness with the prime purpose of improving the foam stability. The nitrogen smooths out the harsh burned character delivered by roasted grains. Furthermore, for reasons that are not understood, nitrogen suppresses the hop aroma of beers. That character is not prevalent on traditional stouts, but a dry hop nose is expected on many a pale ale, and so nitrogen usage in such products (or in lagers intended to display late hop character) is doing a disservice. Equally, the smooth texture imparted by nitrogen is not desirable in beers other than stouts. Again, that is my opinion—one shared by many traditional brewers—but it needs to be stressed that there are some fairly successful beers in the market that are overtly marketed on the concept of "smooth flow" or nitrokeg and that get the characteristic from the employment of nitrogen in the keg and during dispense. In the brewing industry we let customers decide what they do or do not like, as opposed to having self-appointed gurus pontificating about what is good or bad, as with wine.

Imperial Stout

Originally brewed in England for the Baltic market from the late 1700s, it is a strong (9%–10% ABV) dark ale much appreciated in the days of the imperial court of Catherine the Great. Characteristics include intense roast notes and profound bitterness.

Sweet Stout and Other Stouts

Milk (or sweet, or cream) stout was so named for the whey-derived lactose employed in a production process dating back to 1669. Lactose is not metabolized by brewer's yeast, and it has only one-fifth of the sweetness of sucrose. Apart from this modest contribution to sweetness, lactose affords body to a usually lower alcohol (perhaps 3% ABV) product that traditionally errs on the side of caramel/toffee mellowness rather than overt roast/burned flavor. When my wife gave birth to our first child in 1980 in West Sussex, England, she was offered Mackeson, perhaps the most famed of the genre, in the nursing ward.

There are *oatmeal stouts*, with perhaps 5%–20% of the grist comprising rolled oats, which afford a somewhat dry character to the palate. Some people believe that the complex polysaccharides from the oats contribute to a smoothness in this style of beer.

Stout has long been associated with oysters. Primarily this was in the form of stout accompanying a plate of the shellfish, but from time to time there have been *oyster stouts*, with the beer allowed to "rest" on a bed of oyster shells or, in later years, suffused with an oyster flavoring. I recall asking one brewer who used the oyster shell approach what he did with the shells after the beer had been pumped out of the vessel. "Oh, we wash them with sodium hydroxide and then water and use them again." Methinks that successive batches of that oyster stout would have had progressively diminished mollusk character.

Stout mixed with bitter ale (traditionally Guinness and Bass, respectively) yields *black and tan*, while *black velvet* is stout blended with champagne.

Winter Ale

Through history in Britain there was a tradition of consuming lightly hopped, maltier brews after heating (mulling) and perhaps mixing with materials such as roasted apples, eggs, spirits, and spices such as nutmeg. Toasted bread might even be floated on the hot beverage. Latterly in the United States ales flavored with proprietary and confidential blends of spices are unveiled each winter holiday season. A prominent producer of such beers is the Anchor Brewing Company in San Francisco. *Pumpkin ales* might be classified here. There is clear historic precedence for such beers, the early settlers employing pumpkins in lieu of the scarcely available malt. Latter-day pumpkin ales owe more to pumpkin spice than to pumpkin.

German Ale

Alt means "old," and *Altbier* (often abbreviated to *alt*) is "brewed in the old way," as opposed to the production of lagers from the more southerly climes of Germany. The grist is of barley and (to a lesser extent) wheat malts. *Alts* tend to be a dark copper color and relatively bitter, with an ABV value a little under 5%. The style emerged in Düsseldorf, and the temperature for fermentation is somewhere between traditional ale and lager levels. The style is frequently matured for rather longer than would be the case, say, for an English ale.

Kölsch, from Cologne, is a much lighter beer than *Alt*, with the darker malts used in the production of *Alt* replaced by the less intensely kilned malts. It is also fermented at atypically low temperatures, so it is often referred to as an ale-lager hybrid. The brewers of Cologne describe the style as "a light colored, highly fermented, strongly hopped, bright, top fermented Vollbier" (*Vollbier* means a middle-of-the-range product, neither low nor excessively high in alcohol). These traditionalists also prefer that the beer be served in straight-sided six-ounce glasses. Oh, and they eschew *Kölsch* brewed by anyone other than a Cologne-based brewer. Just as for Bordeaux, Champagne, and Chianti in the world of wine, *Kölsch* has been granted appellation status. (*Cream ales* in the United States are comparable to *Kölsch* in some ways.)

The German wheat beers are made with top-fermenting yeast and are thus ales. *Weizenbier* is made from a grist comprising at least 50% wheat malt. (The use of some malted barley in the grist is important, as it provides the husk filter bed, which is not available from wheat malt.) This style tends to be relatively highly carbonated, with pronounced fruity notes but also a distinct clove-like character, both of which derive from the special yeasts used to make such beers. Absent such banana and clove-like notes, a *Weizenbier* is not authentic. Such beers tend to be relatively pale or strawlike in appearance—and, to be bona fide, should strictly *never* be served with a slice of lemon. These beers are most frequently *Hefeweizens* and are cloudy due to the presence of a yeast residue, which is traditionally employed to carbonate the bottled product through natural conditioning. The term *Kristallweizen* denotes that the beers have been filtered to remove yeast.

A related beer style is *Gose*, which tends to be tart, relatively salty, and lemony, and there is also the darker *Dunkelweizen*.

Meanwhile stronger beers of the genre are known as *Weizenbock*, or for still more alcohol one can shoot for *Weizendoppelbock* and *Weizeneisbock*. These are all ales, despite the name *bock*, which generally is ascribed to lagers (see later).

Berliner weisse is much weaker (e.g., 2.8% ABV), very pale, slightly turbid, and made from a grist of less than 50% wheat malt. Lactic acid bacteria are used to generate a low pH of 3.2–3.4 and therefore much sourness. These beers tend to be taken with a dash of raspberry or sweet woodruff (*waldmeister*) syrups. Napoleon Bonaparte called them "Champagne of the North."

Roggenbier is a Bavarian style made with at least 30% malted rye.

Steinbier is a style defined by the production technology: the primitive use of hot rocks to heat up the wort. As ever, there are those who seek to reinvent archaic technology in pursuit of a sales storyline.

Belgian Top-Fermentation Beer

One of the famous genres is *Trappist* beers, which must be brewed within the walls of a Trappist monastery. The brewing members of the International Trappist Association (ITA), which embraces all of the monasteries where Trappist beer is produced, are currently Orval Abbey (Orval, Belgium), Our Lady of Saint-Remy Abbey (Rochefort, Belgium), St. Benedictus

Abbey of De Achelse Kluis (Achel, Belgium), Our Lady of the Sacred Heart Abbey (Westmalle, Belgium), Scourmont Abbey (Chimay, Belgium), Sint-Sixtus Abbey (Westvleteren, Belgium), Our Lady of Koningshoeven Abbey (Tilburg, Netherlands—the beer being labeled La Trappe), Maria Toevlucht Abbey (Zundert, Netherlands), Stift Engelszell Abbey (Engelhartzell, Austria), Mont-des-Cats Abbey (Godewaersvelde, France), Saint Joseph's Abbey (Spencer, Massachusetts, United States), Tre Fontane Abbey (Rome, Italy), Monastery of St. Peter of Cardeña (Burgos, Spain), and Mount Saint Bernard Abbey (Leicestershire, UK).

Trappist products have distinct fruity notes and strengths as high as 12.5% ABV, the high alcohol content being derived in part from the use of a proportion of candi sugar in the kettle. The relative strength of the brews is loosely marked by the terms *enkel* (single), *dubbel, tripel,* and *quadrupel,* though such terms are just as likely to indicate other characteristics of the brews. For example, a *dubbel* is an alcoholic brown ale, whereas a *tripel* is much lighter in color. Beers in the Trappist style that are not brewed in one of the monasteries are termed *abbey ales.*

A very distinctive Belgian beer style is *Lambic.* These products have complex flavor characteristics due to the metabolic activities of diverse microflora beyond brewing yeast alone (and as the brewing strains are primarily *S. cerevisiae* I am including Lambics in this section), a complexity encouraged by the cooling of worts in coolships that would be considered an insufficiently hygienic way to do this by brewers of most other types of beer, who steadfastly adhere to the use of defined brewing strains of yeast alone. Lambic beers tend to be quite sour (low pH) and are frequently not clear ("bright"). They are produced from grists containing significant quantities of unmalted wheat and the hops employed tend to be deliberately aged. It is common to enhance the complexity of such beers still further by adding fruits such as cherries (*kriek*), raspberries (*framboise*), blackcurrants (*cassis*), apples (*pombe*) and peaches (*peche*), and these beers may be aged in used wine barrels. *Gueuze* represents a blend of old and young Lambics conditioned by yeast in bottles typically sealed with a cork and wire-cage.

Belgian red ales represent another beer that does not simply involve brewing yeast in its production, in this instance lactic acid bacteria being employed to deliver tartness, and there is a prolonged ageing in oak barrels. *Oud bruin* is a relative—the *oud* referring to prolonged aging for a year or more.

Saison is a primary segment in the *farmhouse ale* category, so-called for the *saisonaire* migrants who came to work on the farms for the harvest.

Reflecting perhaps the origin of a beer that would have meant different things to different farmers brewing in their own home, it is hard to be firm about what precisely a saison comprises, ditto the close relative from France, *bière de garde*. They tend to be dry, effervescent, and fruity. *Saisons* are more "hop forward" than the maltier *bière de garde* is. In Finland, the equivalent style is *sahti*.

Witbier (in the Flemish, the French version being *bière blanche*) is a straw-colored style of beer, cloudy because of its unfiltered nature and produced from a grist of unmalted wheat, malted barley, and perhaps oats. They frequently incorporate orange peel and coriander in the making, as well as some lactic souring.

Bottom Fermentation Beers

Pilsner

What is perhaps the classic within this style originated in mid-19th-century Bohemia (now the Czech Republic), in the Bürger Brauerei (Citizens Brewery) in Pilsen that had been built in response to severe concerns about the quality of beer in the city hitherto. Indeed, they poured all the beer away in 1838. Brewmeister Josef Groll had learned how to make great malt from studying English approaches to malting, and he spirited yeast (and a few student brewers) away from Bavaria. The result was the original Pilsner. It is quite a malty brew, typically with 4.8%–5.1% ABV and a pale gold color. Particularly prized is the late hop character. Such beers in the Czech Republic frequently display a distinct butterscotch or popcorn flavor due to diacetyl. For the vast majority of beers worldwide this is deemed a severe defect, but it is expected by the Czech drinkers, who consume more beer per head than the folks in any other nation on the planet. Herein is further evidence that there are no hard-and-fast rules about what is and what is not a good beer.

All too often, the term "lager" is used synonymously with Pilsner (Pils). Lager as a term is really an umbrella description for brews fermented by bottom-fermenting yeasts strains, usually at relatively low temperatures. There are many lagers beyond pilsner.

Bock

As a style, bocks tend to be stronger than pilsners, at 6%–8% ABV, and the labels of such products typically portray goats as a symbol of strength. This type of beer originated in Einbeck in Germany in the 14th century. Bocks typically have sulfury and malty flavors with colors that range from straw to dark brown. They are less hop forward than Pilsners. *Doppelbocks*, originally from Munich, may contain up to 12% ABV. They usually have a reddish/brown color.

Other variants include *Maibock*, which are amber-gold in color, and *Winterbocks*, which tend to be darker.

Marzen

"March beer" was traditionally brewed in March. The style originated following the 1553 edict that brewers were forbidden to brew between April 23 (St. George's Day) and September 29 (Michaelmas), some say to avoid increased risk of microbiological spoilage, while others suggest it was to minimize risk of fire. Accordingly, the brews were historically of relatively high strength (up to 6.5% ABV) to allow a long shelf life through the prolonged lagering process. The last of the beer was finished off in the Oktoberfest (formerly in October, but moved to September when the weather is better). There are pale and darker versions depending on the blend of malts employed. The latter are sometimes referred to as *Vienna* lagers. Another style in this general category is *Oktoberfestbier*, a name that in Germany can only be used by the six breweries permitted to sell their products at the festival in Munich: Paulaner, Augustiner, Löwenbräu, Hacker-Pschorr, Spaten, and Hofbräuhaus.

Helles

The German word *hell* means "pale." These pale yellow to amber lagers tend to be malty (though less "full" than a *Marzen*), of relatively low bitterness and hop character, and around 4.5%–5.5% ABV. They are the typical go-to beer in Bavaria.

Dunkel

From the German word for "dark," this style has comparable flavors and strengths to *Helles*, but these beers are copper-brown or mahogany in color through the use of more strongly dried malts.

Schwarzbier

As the name indicates, these are black lagers. They are dry with alcohol contents between 3.8 and 5% ABV. Strongly roasted malts are employed in their production, perhaps from de-husked grain to avoid quite the astringency that is afforded by a regular chocolate or black malt.

Rauchbier

These are hugely popular in Bamberg and are produced from a grist that incorporates malts dried over burning beechwood and thus they have aroma notes comparable to those of the peated malt whiskies but that more commonly are likened to smoky bacon. Any beer style (Marzen, Bock etc.) can be produced with such malt. Some of the smoked character adsorbs to the yeast used to ferment these beers and there are brewers who use the yeast from a Rauchbier brew to pitch other brews not made with smoked malts but to which they seek to introduce a hint of this character.

Kellerbier

"Cellar beer," this is a malty and yeasty unfiltered draft beer drunk from earthenware mugs in the heart of Bavaria. With an ABV in the region of 5.1%, it is amber-colored due to caramel malts and is quite hoppy.

Malt Liquor

Malt liquors in the United States are products of relatively high alcohol content (6%–7.5% ABV) that are very pale, very lightly hopped, and quite sweet.

For the most part, rather than being a style of beer, it is really a labeling issue founded on the identification of stronger beers, with state-to-state variations. The grist will typically contain rather less malt than most other beers and a comparatively large amount of adjuncts such as corn and sugar.

Steam Beer

The origins of this product can be traced to the California Gold Rush and a demand for light and refreshing drinks despite the unavailability of ice for cold storage and conditioning. Bottom-fermenting lager yeasts were used at warmer fermentation temperatures than was customary for lagers, in shallow vessels into which the "steaming" wort was introduced to cool. Others insist that the term "steam" rather relates to the effervescence of the brew as perceived by the consumers of yore as the beer was dispensed in bars. Fritz Maytag and his Anchor Brewing Company obtained labeling rights over the term, leaving other companies to refer to beers of the style under the category of *California common* beers. Just as *Kölsch* is a hybrid (ale yeast working at low temperatures), so too is this style of beer (lager yeast fermenting at high temperatures).

Other Beers

From my youth in England I well recall *shandy*, in which pale ale and lemonade (more 7-Up than mom's own freshly squeezed stuff) were mixed in equal quantities. Such drinks are extremely refreshing after playing a sport. The French have their *panache* and the Germans *Radler*. Some folks used to add a dash of blackcurrant juice to their lagers, or a splash of lemonade ("lager top"). *Russenmass* is *Hefeweizen* with lemonade, while *diesel* is a lager blended with a cola.

A well-known English brewer developed a famous *chocolate stout*, in which substantial chocolate malt is used, but also a bar of chocolate and some chocolate-flavored essence. Other chocolate stouts and porters owe their character to chocolate malts and perhaps cocoa nibs.

Chilies, bacon, Rocky Mountain oysters, squid ink, civet shit, goat brain, stag semen, stale bread, baseball bat chippings, fried chicken, vaginal emissions, and roast lamb . . . there is no limit to what brewers have used in recent times. This is a transparent marketing strategy of "Look at me and what I have done. Isn't that outrageous?" I recall a recent sampling of a beer (from a huge company—it is not only the smaller brewers that are pushing the envelope) containing ginseng, caffeine, and guarana. As my irises dilated, the chap who had offered it to me smirked and said, "It wasn't designed for you, Charlie." This was a party beer. Like many of these concoctions, thankfully, it was a gimmick that did not last long in the production schedule. There really is something to be said for the Reinheitsgebot, the so-called German purity law, introduced by the Duke of Bavaria in 1516 and which restricts the materials that can be used to make beer to malt, hops, yeast, and water. Yeast wasn't originally listed because, amazingly, nobody was agreed on its existence until the mid to late 19th century. This law has become something of a holier-than-thou belief set, with its adherents being as insistent that it is the only route to decent beer as the folks at CAMRA have been insistent than only cask ales have any merit. It's ironic to remember that the law was introduced specifically to stop the prior practice of using decidedly dodgy materials, such as cow bile. It is my belief that it is rather too restrictive—there are some great beers made with the aid of adjuncts, not least the Trappist beers we discussed earlier. However, anything to outlaw the use of blatantly ridiculous ingredients in beer does rather have merit in my eyes. I really am not a fan of peanut butter, carrots, or coffee beans that have emerged from the anus of an elephant finding their way into beers. I spoke at a meeting in London one time on this issue and was referred to as a Luddite. If that means I deplore stupidity, then I will own the name tag.

There are several other traditions in the world of mixing beer with things that seem rather more tolerable (it's all judgmental after all). One of the best known is *michelada* in Mexico, in which lager is mixed with lime juice and salt. Sometimes Worcestershire sauce, Tabasco, or clam juice is used, but then the product is not specifically *michelada*.

A genre on the cusp between beer and wine comes in the form of the *malternatives*, sometimes known as flavored alcoholic beverages or alcopops. They are marginally hopped beers, with the color removed and a diversity of flavors introduced. They are produced in this way for tax purposes—being beer-based and with a strength typical of beers, they are taxed as such in many states, this being at a lower rate than spirits. There is a strong lobby to levy the higher tax on them, for it is believed in many quarters that these relatively sweet products are targeted at the younger element, those who have not yet acquired a taste for bitter products. Prominent brand names include

Smirnoff Ice, Mike's Hard Lemonade, and Hooper's Hooch. I, for one, would tax these far more heavily than regular beer. By regular I mean those produced with traditional techniques and with traditional materials (and the latter includes a diversity of time-honored grist materials that I have drawn attention to already).

Ice Beer

The Germans have been brewing *Eisbock* for many a long year. In the process, the beer is taken to below the freezing point, which brings water out as ice, but leaving the alcohol in solution as ethanol has a lower freezing point than water. Thus, the strength of the beer is increased, to a minimum of 7% ABV.

The ice beer story in North America, by contrast, is an example of how a beer style emerged through the use of a technology that was originally installed for another purpose. In the 1980s, many brewers worldwide had decided that, rather than ship finished beer to its destination, it would make economic sense to transport the beer in a concentrated form and then reconstitute it at the point of sale. They experimented with "freeze concentration," which is basically the *Eisbock* technique. Most of the beer components remain in solution in a concentrated form.

Labatt, a major Canadian brewer, was one company that experimented with the technique but discarded it because it would contravene certain legislation that dictated that beer had to be brewed in the province in which it was sold. Fortunately for my close friend Graham Stewart, Labatt's technical director at the time, and his colleagues, they hit upon a very different use for the equipment. They were looking for a new angle on beer marketing and identified *ice* as being a powerful concept that associated extremely well with beer in the perception of Canadian drinkers. As Professor Stewart says: "After all, Canadians already knew all about putting beer out onto the window ledge in the winter, freezing ice out from it, thereby increasing the alcohol content." By the early 1990s a new and exciting beer story was being told. and most major brewers developed their own ice brands. In truth, this was not really an exercise in producing substantially more alcoholic brews: in the process there was the merest suggestion of ice production, just to allow a story to be told.

Dry Beer

The mid 1980s saw the emergence of dry beer, and through it the astonishing growth of the Japanese brewer Asahi. It launched a new brand called Super Dry and saw a 25% increase in market share within three years. As the name suggests, it is a straightforward concept analogous to dry wine: a lager with a relatively low proportion of residual sugar, but clever marketing, and the characteristic outstanding package quality associated with Japanese brewers made it a clear winner. It was a product deliberately designed to appeal to as many people as possible through having no extreme flavor characteristics that might alienate sections of the populace. In no time it was followed by other dry beers ("me toos"), and a dozen countries contributed over 30 new brands of dry beer.

Light Beer

We started to chat about these earlier. Premium light beers now constitute the most popular beer category in the United States and have come a long way from the first reduced-calorie brew by Rheingold. These beer styles are differentiated by their content of residual carbohydrate: standard premium beers contain a proportion of carbohydrate (dextrins) that survives the fermentation process, whereas a light beer has most or all of this material removed. Therefore these beers have lower calorie contents, provided they do not contain extra alcohol, which in itself is the main contributor to calorie intake in any alcoholic beverage. The Food and Drug Administration stipulates that such beers should contain a third fewer calories than the regular equivalent beer.

Such beers might be made by simple dilution or by tweaks in the mashing stage that allow a greater proportion of the starch to be broken down. The easiest way is to add an extra enzyme to the mash or fermenter that allows the dextrins to be converted to glucose, enabling all the starch to be converted into alcohol, leaving no remaining carbohydrate. We referred to this enzyme earlier in the context of brut IPAs. Of course, this means a higher alcohol content, so to avoid those extra calories in light beers the liquid is diluted to an even lower final ABV than the equivalent "regular" beer. For instance, Coors is 5% ABV, whereas Coors Light is 4.2% ABV. The extra proportion of water makes these products thinner than those that have their dextrins surviving.

The biggest growing brand in the sector is actually marketed overtly as a low-carb beer and is even lighter (character-wise) than the company's well-known light brand. This author has a real problem with labeling beers as low carb, on the basis that it insinuates that all other beers are "high carb." They generally are not, as most of the sugar is usually converted into alcohol. Most of the carbohydrates in beer are bigger than sugars and are either non-metabolizable in the human or constitute "slow release" carbohydrates, which are more desirable than those that boost the glucose content of the bloodstream rapidly.

Draft Beer

The word "draft" ("draught" in the British Isles) can refer to two entirely distinct beer types. Traditionally it refers to beer dispensed from kegs or casks via pipes and pumps, or indeed straight from the cask located at the back of the bar, as is still the case for some of the traditional English ales. It is also used, however, to describe bottled or canned beer that has not been pasteurized but rather sterile-filtered. The marketers had a new angle for this small-pack beer: "as nature intended." Much beer worldwide is now marketed on this angle of "non-heat treated." For instance, no beer in Japan is pasteurized. The fact is that, provided the oxygen levels in the beer are low prior to pasteurization, this process has no adverse impact on flavor and is actually to the benefit of foam, which can deteriorate with time in non-pasteurized beers. It certainly is a curiosity that the lack of pasteurization is taken by some as a positive thing. I cannot imagine most customers caring much for this if we were talking milk.

From Cask-Conditioned to Nitrogenated Beers

The emergence of nitrogenated keg beers ("nitrokegs") in the United Kingdom is an informative lesson in how modern technology can throw up products whose origins are in traditional practice.

The classic beer style in England is non-pasteurized ale of relatively low carbon dioxide content. Happily, many famous brews of this type continue to be produced, as championed by CAMRA. The production of traditional English ales involves their going from fermentation into casks, to which are added hops, sugar, and finings materials that help the residual yeast to settle out. That yeast uses the sugar to carry out a secondary fermentation, which carbonates the beer to a modest extent. The product is not pasteurized and must be consumed within a few weeks. It is characterized by a robust, hoppy flavor but also by much less gas "fizz" than other product types. Hops may be added ("dry hopping"), but there is a risk because hops tend not to be micro-biologically sterile.

Again in the mid-1980s and with the projected demographic shift to more drinking at home rather than in pubs and bars, marketers in the British Isles decided that they would really like to be able to sell this type of beer in cans for domestic consumption. The problem had to do with the low CO_2 content, for the gas is generally required to pressurize and provide rigidity to cans, and also to put a head on beer. For cask beers, it is the hand pump characteristic of the English pub that does the work in frothing the beer. For "normal" canned beers, the relatively high gas content does the job for you on pouring. So how could the foaming problem be overcome for canned beers containing relatively little carbon dioxide? The answer was the "widget," a piece of plastic put into the can that flexes when the can is opened and causes bubbles to come out of solution (see chapter 5). This technology had been invented by Guinness, a brewer with a long tradition of producing stouts with superbly stable heads. Allied to this was the realization that nitrogen gas makes vastly more stable foams than does carbon dioxide, again a technology that had been pioneered by the Irish company and taken advantage of by many brewers to enhance the heads on their draft beers. Therefore, nitrogen was included in the cans—dropped in at canning in its liquid form. Not only does the nitrogen help the foam, but it also brings a smoothness to the palate, enhancing the drinkability of some of the beers that contain it, notably the stouts. The sales of canned beer with widgets zoomed, and, seeing this, brewers recognized the potential for so-called nitrokeg beers, where the beer is on draft dispense, but is characterized by low CO_2 and the presence of N_2.

Every brewer did not appreciate the merits of nitrogen and widgets. Tired of complaints about the taste of the canned beer having taken a turn for the worse since the introduction of the "lump of plastic," one English company reversed matters, eliminated the device, and proudly announced on the label "widget-free ale." Applause from this author, for one. Another company, eschewing the widget, simply put the words "For head pour quickly" on the can.

Non- and Low-Alcohol Beers

"Normal" beers range in their alcohol content from 2.5% to 13%, with a handful of extreme products (read on). To a Bavarian used to beers having 6% alcohol or more, the regular tipple of the English ale drinker at, say, 3.8% might be viewed as "low alcohol." Non- and low-alcohol beers (NABs/LABs) can be classified in many ways. For our purposes I will define them as beers containing less than 0.05% and less than 2% alcohol (by volume) respectively.

While there are a few successful NABs/LABs in the world, they are the exception rather than the rule. For many people it is a contradiction in terms to associate a beer with low alcohol: after all, what is a beer if it does not deliver a "kick," albeit with less "bang" than the more alcoholic wines and especially spirits? The rationale behind developing such beers in the first place is an interesting one, and is largely based on the proposal that peer pressure among drinkers convinces some people of the need to be seen to be drinking a product indistinguishable (by sight) from a normal beer, but one that is of reduced alcohol content, thereby enabling the imbiber to drive. Increasingly, it has been appreciated that this peer pressure phenomenon was overstated and that educated consumers will happily drink something that nobody associated with booze, say a juice or a cola, if the circumstances demand it. It seems that the only justification for purchasing a beer of low alcohol content is if it is pleasing to the palate. Moreover, that certainly has not always been the case for many such beers. There are those who tell me that the quality of such products is getting better. For my part, I have a psychological block precluding them from passing down my gullet. Just like decaffeinated coffee. Why bother, unless they are genuinely delicious?

One thing I did do in my more sporty days when I was looking to rehydrate with a post-playing drink without overdoing things was to mix together half a pint of Carling Black Label (4% ABV) with half a pint of its alcohol-free equivalent, Barbican, to produce a pint at a strength of 2% ABV in which the positives of the former certainly masked the negatives of the latter. We never marketed such a product outright but did sell a beer that was essentially put together in the brewery by blending one part lager with three parts of dealcoholized lager, to render a 1% ABV brew. It was not as palatable in my view as the beer that I mixed for myself in the bar, but at under 1.2% ABV it was not taxed. A 2% ABV product would have been taxed and thus much more expensive.

Alcohol-free and low-alcohol beers have been made in many ways. Perhaps the most common techniques are either limiting alcohol formation in fermentation or stripping out the alcohol from a "normal" beer. In the first case, the yeast can be removed from the fermenting mixture early on, or indeed the wort that the yeast is furnished with may be produced such that its sugars are much less fermentable, or a yeast producing little alcohol might be tried. Alcohol can be removed by reverse osmosis or by evaporating off the alcohol using vacuum distillation. It should come as no surprise that attempts to remove alcohol will also result in the stripping away of desirable flavors. Equally, if fermentation is not allowed to proceed to completion, these very flavor compounds are not properly developed, and undesirable components derived from malt are not removed. Either way, the flavor will be a problem. And considering that ethanol itself influences the aroma delivery of other components of beer, as well as itself contributing to flavor, it will be realized why good NABs/LABs are few and far between.

The Other Extreme

There is a war going on for who can come up with the most alcoholic brew. It's another case of "Hey, look at us," and can hardly be said to be founded on a basis of responsibility and finesse.

It started with the Boston Beer Company and a product called Utopias. The tenth iteration came in at 28% ABV and retailed at over $200 a bottle. It's a pretty looking package but . . .

Over to Europe and those habitual rivals, the Germans and the British. Cue Schorschbräu from a town between Nuremberg and Stuttgart, who knew all about *Eisbier* technology and the habits of those Canadians sticking their booze in the snow to freeze out water and boost the alcohol. Result: Schorschbock 31, that is, 31% ABV.

Thence to Scotland and those naughty boys from BrewDog. (I can't help but like them—we at the university featured on their television series in the United States. And there is no doubt that Martin Dickie and James Watt know their stuff. They are still impish, though.) They came out with Tactical Nuclear Penguin. Tactical to attack the Germans. Nuclear: 32% ABV. Penguin: all that ice, get it?

Back to Germany, where no doubt the folks were saying *Scheissen, Gott in Himmel*, and the like, before they came up with Schorschbock 40, 40% ABV. *Ja, zat vill show zem.*

Nothing to the lads from Ellon, near Aberdeen in the rugged north of Scotland. They produced a beer of 41% ABV and called it Sink the Bismarck. Hardly Entente Cordiale—or *Freundliche Vereinbarung*, you might say.

Then the Scottish lads surpassed themselves with the End of History, which I guess is what it might be if you drink too much of it, 55% ABV and every bottle encased by a taxidermist's best. Squirrels and stoats. Bottles can sell for $20,000. The title comes from the work of the philosopher Fukuyama. Hmm.

Remarkably, it is not the strongest beer anymore. The world record is currently with Snake Venom, weighing in at 67.5% ABV, from Brewmeister, a brewery in Keith, Scotland.

I ask myself, "Why?" I guess the retort would be like that of George Mallory tilting at Everest: "Because it is there."

4

Shopping

I am not exactly sure where it was, but most definitely it was on or near Route 395 as we headed from Disneyland to Mammoth Lakes, the latter more to my personal preference than the former. (Like anything else, including beer, it's all a matter of taste.) We were on a vacation sometime before a move to live in the States was ever contemplated. We stopped for gas and I was intrigued to find that they also sold beer, bait, and bullets. This seemed mightily strange to an Englishman.

In those days the beer selection was modest. Chances were that there was Bud, Miller, and Coors and light/ice/dry/whatever versions thereof. Which of itself is okay—there is certainly nothing wrong with these beers, enjoyed by millions. My point is that now, a quarter of a century later, there are many more products to choose from—but how do you know if they are any good or not?

Good is, of itself, not easy to define. What is good to you is not necessarily good to me. It is like anything else in this world. To my wife, Marmite is good. To me it is evil incarnate in the shape of icky black sludge. Good to you might be rap, maybe Eminem or Lil Wayne, and you may very well not share my passion for Satie and Ravel. My idea of good art is Constable. You might opt for something that to me resembles something that an overwrought orangutan might have generated by smearing assorted paints onto a blank canvas.

For any product, good is surely "true to expectation." Therefore, focusing solely on beer, your purchase should match your anticipation. Do you recognize the qualities of the beer as what you forecast them to be?

However, this leads us headlong into the real challenge in these days when brands are available further and further away from home base. Let me illustrate.

Consider an aficionado of Pilsner Urquell (the latter word meaning "original source" in the Czech Republic, where they drink more beer per capita than any other country on the planet). He or she is expecting the beautiful golden color, lashings of foam—and a distinctive buttery or popcorn nose. The latter is due to modest but unmistakable amounts of a chemical called diacetyl in the product, a substance that for most beers and most brewers represents an unforgivable flaw. Pilsner Urquell, though, is supposed to have it. Now ship that beer a few thousand miles to Sacramento, across who knows what temperature challenges and some less than gentle yawing on board ship. This journey by way of truck to a port, ship, more trucking, and usually indeterminate and ill-regulated storage in warehouses before it reaches the store or the restaurant will have nicely aged the product. "Nicely" is an ironic term. When beer ages it develops a myriad of aged characters, of which tomcat pee and wet paper are two of the most prominent. And so this beer will likely taste like feline micturition outpourings and sucking on a cardboard box. Yet this is what customers in the United States are encountering when they purchase this beer. They *think* it's supposed to taste that way. The untrained taster is not predisposed to thinking, "Oh my, this is catty and, golly gosh, I get the feeling that I am chewing on damp parchment." They just accept it for what it is. Should they journey to Prague and taste the fresh product, they will get a shock. Likewise should Czechians land in California and eagerly seek out a Pilsner Urquell, they will likely be somewhat distressed.

We have a problem. The reality is that (as I write) some 34.5 million barrels, or 17.5% of the total beer consumed in the United States, is imported and I would venture to suggest that at the very least half of it is subpar in flavor, if we are to compare its taste and smell with what the brewer intended. (In passing, let me also state that the so-called craft brewers produce 12.7% of the nation's brews.) Commercial expediency in the shape of the desire to sell brands over bigger and bigger distances outstrips the technical ability to prevent flavor change. Despite years of extensive and sophisticated research into the problem, the stark reality is that beer flavor will change. There are somewhere between 1,000 and 2,000 different chemical species derived from malt, hops, water, yeast, and whatever else finds its way into the production of beer that contribute to aroma and taste (probably twice as many as feature in wine) and a detectable shift in the amount of any of them (to higher levels or, indeed, lower levels) will take the flavor away from what it should be. It is frankly impossible to prevent flavor change.

For some beers you might legitimately "talk up" flavor change. Notably I am talking about the higher-alcohol beers, such as barley wines, which (it can be argued) mature in interesting and beneficial ways. Nevertheless, in all honesty most beers, like a good many (but not all) rockers, do not age well.

So let us take this unfortunate scenario with us into the store as we decide which beers to choose, rejoicing as we do in an era when there has never previously been a better selection of products than there is right now.

The first thing you should be looking for is a refrigerator (Fig. 4.1). To understand why, let me tell you that there are three general things that can be done to keep beer fresh for as long as possible, a trio of actions that will slow down the change in levels of many of those 1,000–2,000 substances I just spoke of.

The first is that the oxygen level in the beer should be as low as possible. These days there are superb packaging machines for filling beer into bottles and cans that allow the brewer to get extremely low levels of air in the package. Oxygen is a terribly bad influence when it comes to beer flavor because it reacts with several types of molecule in the product to convert them into those undesirable smells and tastes mentioned earlier.

Let me take this opportunity in this context to explain that this is the reason why beer will have a longer shelf life in a can than it will in a bottle. In the actual filling of a can you cannot hit quite such low oxygen levels as you can in bottling (though the difference is not *that* huge). However, that is it: no more air is going to get into the can. With bottles, however, air sneaks in between the neck and the cap (called a crown cork). So beer goes stale faster in a bottle than in a can. The tighter the crown cork is applied, the better—so if you are opting for bottles, then best to go for ones that need a bottle opener to pry off the lid and not the ones that you can twist off. (No simple way of checking—don't try opening them in the store, because if they are not twist-off, you will get sore fingers; if they are twist-off, you can't twist them back on and the storekeeper is going to be displeased.)

The other thing about cans is that no light can get into them. This is not the case for beer in glass. The light reacts with the hop-derived substances that give beer its bitterness, and they break down to give the unmistakable whiff that is closely similar to that which is emitted from between the hind legs of a skunk. Brown glass is reasonably protective, but green glass and clear glass are not. So that is why buying beer packaged in that color of glass is a risk. Even the light used to illuminate the beer in the store is enough to turn the product skunky. Now some beers have for the longest time been made what you might call skunk proof, by using chemically modified hops. Miller was the pioneer. The hops are dried, pulverized, extracted in liquefied carbon dioxide

Fig. 4.1. Refrigerated beer in the store: (a) a closed-door cooler, (b) an open-access cooled cabinet. Photos by the author.

and chemically adjusted to a form that is even bitterer but that is no longer sensitive to light. Many brewers are not prepared to use such materials. As a consumer, you make the choice. If you do not feel strongly about it, then no worries: the beer is perfectly safe. Equally, about a third of customers prefer the skunky version of a famous beer packaged in green glass. And that is fine, too. If, however, you do not care to consume something with this aroma and likewise you are not enamored by the idea of transformed forms of bitterness, then you had best choose beer in brown glass (or a can).

There is another advantage to cans: they are lighter than bottles and they can also pack into smaller spaces. However, there is a psychological factor at play. I was in a restaurant in Oklahoma City one time with my wife Diane and younger daughter Emily. I knew it was a nice restaurant because the table had a crisply starched white cloth on it. The charming waitress pitched up and took our drinks order.

"What craft brews do you have?" I asked. (See, even I lapse into using the term.)

"Well, I've got Bud, Miller, and Coors," was the reply.

"Is that it?" said I.

"Nowp, I got one more," she said.

"What's that?" I wondered.

"Pabst Blue Ribbon," the waitress replied.

"I'll have one of those," I said. And along it came. The good news was there was a glass. However, the beer was in a can, upon seeing which I winced.

"What's the matter with you?" said my wife. "You stand up and pontificate about beer being fresher in a can. They bring you a can and you moan!"

"But, my dear, a can does not look classy in a restaurant!" Think about it. Did you ever get canned or boxed wine in a restaurant?

Yep, I know, it's stupid of me. But this prejudice is shared by many people. There is one belief set, however, that many folks have, which is that beer from a can tastes metallic. It does not. The inside of cans and lids are coated, and there is no metal pickup on pouring or, indeed, in storing the beer.

Back to those three general solutions to lessening the rate of beer change: let us turn to sulfur dioxide. This material is pretty darned good at lessening the stale character in beer. The problem is that in the United States you have to mention its presence on the label if there is anything like the amount that is likely to do any good. As I write, I am not aware of any beers across the 50 states that have the words "Contains sulfites" on the label. There cannot be many wines that do not. Moral high ground for beer right there.

The third precaution, which is where we entered into this discourse, is refrigeration. The sad truth is that beer does not like heat. Let me spell out the facts. Let us say we freshly package a beer in a brown glass bottle and hold it at "classic" room temperature, which is 20°C (68°F). We would find that it would get to a stage where we are recognizing aged character in about three months. Let us increase that temperature to 30°C (86°F). We will find that those aged characters are appearing after one month. Now let us shoot to 40°C (104°F), which it quite comfortably (or uncomfortably) is in my garage in Davis for weeks on end in the summer months. The beer is stale in just over a week. However, if we take the temperature down to refrigerator temperatures (around 4°C or 39°F), then we can extend the shelf life dramatically, perhaps to as long as a year.

Small wonder, then, that I simply will not buy beer in a store that does not have that beer in a refrigerator. Now straightway those companies that do just stick the beer on the shelf (Fig. 4.2) will argue that the beer is turning over rapidly and that it does not linger on those shelves for long at all. Understood—but it is but one stage in the journey from packaging in the

Fig. 4.2. Beer on a store shelf. Note that most of the brands are imported. Photo by the author.

brewery to being opened and enjoyed (we hope) by the customer. It should be kept cold throughout (although, as we will see elsewhere, purists may wish to take the beer to the appropriate temperature for its enjoyment just before they drink it). It is the reason why responsible brewers, with Sierra Nevada a shining light, insist on cold handling throughout distribution and warehousing. It behooves everyone, including the retailer, to respect the beer and thereby respect the customer.

Next: look for clues as to where the beer is brewed. The simple rule of thumb is that beer will be freshest and truest to type when it is brewed close to the purchaser (as well as having been brewed as recently as possible). In relation to this, we find one of the reasons why Sierra Nevada built a second brewery in North Carolina. Thereby it can satisfy the eastern half of the country with the freshest offerings (remembering that Sierra Nevada distributes cold). The other main reason, of course, is that most beers are at least 90% water: it does not make sense to ship heavy bottles, cans, or kegs of beer unnecessary distances. The likes of AB Inbev have breweries across the nation (indeed, world) for this very reason. Remember, too, that cans are lighter than bottles and therefore less expensive to ship.

It is not unequivocally the case that the beer from the brewery up the road is better than stuff that has traveled much greater distances. The brutal reality is that some of the newer generation of brewers churn out appalling beers. So I would modify the recommendation to say, "Purchase trusted brands from conscientious breweries and if they are local, so much the better."

Chances are that you won't find it, but one very useful item on a label is a best-before date or, even better, a filled-on or born-on date, the latter indicating when the beer was packaged. This tells you how old the beer is. Knowing the filling date is the more valuable simply because a best-before date usually does not tell you what the brewer is decreeing as being the acceptable age of the beer. Is it three months, or six months or, as in the case of the United Kingdom, nine months?

To illustrate this let me go back to my beloved mother country of England. The norm there for the longest time is a shelf life of nine months, so the best-before date is three-quarters of a year after the filling date. Apart from those very strong beers that may age advantageously, I can guarantee you that the bulk of the beers will not be at peak flavor performance by the time they come to the end of their shelf life. Therefore, if you are buying a beer in a supermarket in the United Kingdom in June and the best-before date shows

September, then chances are that the beer inside that package is six months old. It will not be fresh.

Again, let me remind you that many a beer will have traveled a long way to get to you. It might have traveled thousands of miles in unrefrigerated containers, abused by heat and turbulence (that is not good if you think of air increasingly squeezing through that narrow gap between the glass and the crown cork).

Which leads me to the question of taking the beer away from the store. Must I keep it cold? Is it a problem if it warms up before I get the chance to get it into my refrigerator at home? The answer is, or rather another question is, where is home? If you are buying the beer in a store in San Diego, California, and are going to drive real slow to Boston, Massachusetts, with it sitting in the trunk, then, not good. However, if it is your local store, then (provided you keep the product in the shade) it will be just fine by the time you get home. Just get it into your refrigerator as soon as possible.

Above all, buy no more than the amount you need and store it cold (though never frozen). Don't be thinking, "Wow, that's a great price" and buy ludicrous quantities of the stuff that will take you an inordinate amount of time to drink.

And if you like stale imported beer or lightstruck beer with the distinct whiff of *Mephitis mephitis*, well, that is okay too. Just do not make me drink it.

Remember that there is a distinction between *storage* temperature and *drinking* temperature. It's only some beers that are best quaffed straight from the refrigerator. Most should be a tad warmer than that.

When folks hear my English accent (carefully preserved after all these years in the States), they always wonder why I drink my beer warm. To which my reply is, do you think that 55°F (13°C) is warm? If so, you have a different idea of warm than I have. But that is the classic drinking temperature for English cask ale, because it is a typical average cellar temperature in Blighty.

The simple rule of thumb is this: the less flavor intensity, the lower the temperature. So a Bud is sublime straight from the refrigerator. A barley wine, though? Best quaffed somewhat warmer, maybe 60°F (15.5°C).

5

Bars

A couple of years ago I had been a reasonably stressful day in London, culminating in a lengthy Underground ride and a bouncy trip atop a London bus. Number 222. The temptation to pop into The Plough before making for my hotel was rather too great. With anticipation, I ordered a pint of a favorite bitter and settled down with my newspaper, *The Guardian*, and cask-conditioned ale (both in the top half dozen things that I miss from the motherland now that I have lived in the States for more than twenty years).

I should have dangled my nostrils in it sooner. However, all too quickly the awful reality hit me. The dreaded whiff of popcorn, butterscotch if you will. A sip confirmed the worst, as the milky sourness assaulted my palate. My beer was contaminated with lactic acid bacteria.

Now, had Mrs. Bamforth been present I would have been hectored into marching the beer right back to the bar. Seldom, however, do I make a fuss. So there was I, supposedly a well-known chap in beer circles, sitting there inoffensively consuming the offensive. I am not sure whether the bigger mug was the pint glass or yours truly.

Of course, I failed to apply my time-honored test for which traditional draught beer to select. I simply ask, "Which of these beers do you sell the most of?" and will choose that, assuming of course that it is a brand that I know or that I suspect I will like. On this occasion, however, there were only two tap handles there and the query would surely have been futile. With the amount of trade this boozer does in the London suburbs, I *ought to* be confident that the beer was not sitting around in the cask, rapidly turning to acid. The failed assumption clearly was that these people knew how to maintain the strictest hygiene in their beer lines. They obviously did not.

I repeat, then, that I failed my fellow drinkers. Perhaps it was because of, rather than despite, my status as a professor of brewing science that I did not take the beer right back, with a lecture on how to look after the most sensitive beer style on the planet and how to ensure that the pub delights the customer with a drink that I have often likened to an angel weeping on the tongue. I did not want to come the high and mighty—but would that not have been preferable to dutifully sipping the sourness to completion and then skulking out of the premises with not a word of discontent? Surely I was letting down all the poor souls who would go on to make the same purchase later that day and perhaps even later?

Cask Ale

Let us therefore take an in-depth look at cask-conditioned ales here and now, with hopes that my readers who seek the perfect pint but who have the bottle, as it were, to complain if their ale falls short of that ideal will have sufficient ammunition to justify their disgruntlement.

Things are pretty much "typical" for making a cask ale in the brewery all the way up to and including fermentation. It is then that the skill (or is it genius? Maybe beauty? Certainly charm) first steps forth. The fermented ale flows into a so-called racking tank and is mixed with two major additions: a little "priming" sugar and a solution of isinglass finings, a material that induces yeast to settle, thereby clarifying the beer.

The yeast left behind after fermentation will take that sugar in the barrel and convert it into carbon dioxide, thereby boosting (albeit to modest levels) the tingle factor in the product. It is called "conditioning." For a cask ale, it is likely to amount to a little more than one milliliter of carbon dioxide dissolved in each milliliter of beer, whereas for a bottled or canned beer and even beer in kegs it is likely to be substantially more than twice that amount. So cask ales are only mildly carbonated and not especially effervescent. That is why you need to put in some real effort using a beer engine to generate a head on them.

Quite who first came up with the isinglass idea is lost in the myriads of time. Isinglass is a protein called collagen. Stroke your cheek (or somebody else's if they are up for it) and you are touching collagen. It is what skin is made from. The collagen from which isinglass is made is not from a human, you will delight in hearing, but rather from the swim bladder of fish, especially those from the sturgeon family cruising the depths of the South China seas (Fig. 5.1). These swim bladders, which not all fish have, are buoyancy aids that the fish fills with air if it wants to float higher, or that are emptied of gas if it wants to sink. In this way the beast can regulate at what depth it floats around in the brine.

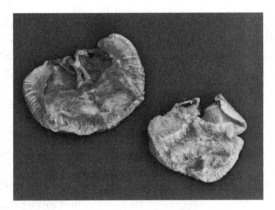

Fig. 5.1. Swim bladders (also known as maws). Thanks to Murphy and Son.

So who first came up with this stuff as a clarifier? One assumes it must have been somebody who thought it would be desperately funny to see how much liquid could be fit into the sac, much as when I, as a cheeky student, rejoiced in filling a condom with water and dangling it out of an upper-story window. The ancient was, however, a brewer (or possibly a winemaker, because this stuff is even more widely used for clarifying the fermented grape) and so he or she filled the convenient little bag with booze. Days later we presume that the individual delighted to see a much clearer liquid emerging from the bag when it was poured out. Eureka, or some such expression of awe will have been uttered—and generation after generation of British ale drinkers have been suitably grateful.

The isinglass as used these days has undergone a fair bit of processing before it finds its way into the beer. The excised swim bladder (the rest of the fish hits the food chain) is dried and diced and then dissolved in a weak solution of acid. The neutralized protein fraction is then ready to be added to the beer (Fig. 5.2).

To be fair, the use of isinglass remains a fairly inexact art form, with little firm scientific information available to help the brewer control its use. We know that swim bladders from different fish (some are small, some much larger) perform in different ways. However, the amount to add is very much a case of trial and error. Most brewers will set up tubes or even glass-ended casks that they fill with newly fermented ale and make different additions of isinglass—one pint per barrel, two pints per barrel, three pints per barrel, and so on (Fig. 5.3). By viewing the ease with which the beer clarifies, the brewer can conclude how much isinglass needs to be added.

Fig. 5.2. Adding isinglass to a cask. Thanks to Murphy and Son.

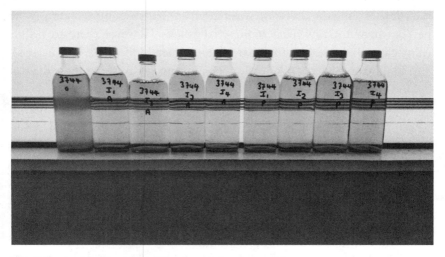

Fig. 5.3. An isinglass trial. Thanks to Murphy and Son.

However, it is not so simple. The beer in the trade will need to display this settling behavior at least twice, because it will be disturbed as it merrily trundles along the highways and byways on the back of a truck and when it is person-handled into the cellar. Then it will need to resettle once set up at the appropriate angle in the cellar (or sometimes behind the bar in the most old-fashioned of establishments). Moreover, beware fluffy bottoms. This is when the settled yeast continues to convert sugars into bubbles of carbon dioxide that start to rise from the base of the vessel and—puff!—up comes strands of deposit which, absent resettling, will very likely enter into the beer drinker's glass.

Straightway you can see why cask ale is far from being an easy option. And I have not mentioned the importance of ensuring that just the right amount of yeast remains in the beer to effect the carbonation of the product.

Back to the racking tank then. Once upon a time brewers used to add another key ingredient at this stage, in the shape of a handful of whole hop cones to give a subtle dry hop note to the product. This practice is far less common these days, for those hops will also likely deliver a charge of undesirable microorganisms to the brew—and remember that cask ale is living beer, with no attempt to eliminate organisms by the agency of tools like pasteurization. This means that cleanliness is next to godliness in the production of this style of product. There again, such is the case for the majority of beers.

When the contents of the racking tank are suitably mixed, then it is time to rack the beer into the casks. These can be of various sizes, most famously the barrel, which in UK terms is 36 imperial gallons. However, there are firkins (nine gallons), kilderkins (half barrel), a hogshead (one and a half barrels), and a butt (three barrels). Actually, the only butts in breweries these days sit behind computers.

The cask has two holes in it, each sealed prior to filling, historically with wood but nowadays with plastic (Fig. 5.4). The one on the circular end is where the tap is hammered in (hence "tapping the barrel"), either a tap literally, for beer to be served directly from the barrel or the tap that is linked to the dispense line that snakes up to the beer engine on the bar. The other hole, filled with a piece of wood or plastic known as the shive, is on the curved surface of the container. It is through this that the vessel is washed and filled and into a hole at the center of which the "spile" is introduced (Fig. 5.5). This wooden peg is used to seal the vessel. At first it is a hard version that ensures that the carbon dioxide produced in the container remains therein. However, if, later, there is a need to lower the carbonation somewhat, then it can be replaced with a softer, more permeable spile. See what I mean about skill—inserting the correct spile at the right time. Just to add to the myriad of things that the brewer and then the bar/cellar staff need to keep in mind.

Cask ale is not pasteurized or filtered in any way. For this very reason, the beer is given a very short shelf life, traditionally 42 days. Once it has been tapped, then we are talking three days maximum. Otherwise we will very likely start to make vinegar. Once the cask is broached, air can start to creep

Fig. 5.4. Casks. (a) You can see in the cask to the front of the operator the location through which the cask is emptied at the top of the cask and the location through which the cask is filled and "breathes" on the side of the cask. It will be realized that on this part of the production line the casks are vertical but when set up in the pub cellar they are at an angle slightly off horizontal—see photo (b). Photo courtesy of St Austell Brewery.

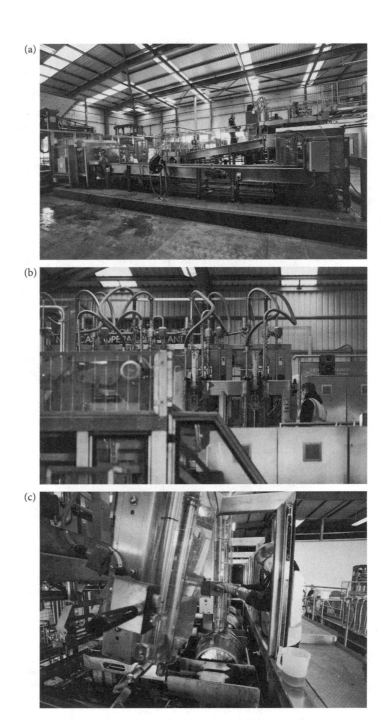

Fig. 5.5. Cask filling: (a) the cask filling facility, (b) the racking heads, (c) filling a cask. All photos courtesy of St Austell Brewery.

in and thus allow for the growth of bacteria that need oxygen, in this instance acetic acid bacteria, the very beasts that make vinegar. Even if there is little oxygen, there are organisms lurking that can grow, making all manner of unpleasant whiffs and tastes. In the case of The Plough, my suspicion is *Pediococcus*, which makes lactic acid, which sours the beer, and diacetyl, the latter giving the noxious stench of popcorn and butterscotch.

Best to go, then, to a boozer where they know how to handle their cask ale. Places that do not leave the barrels on tap for more than three days. Places that clean out the lines comprehensively, before repriming those lines properly with fresh beer.

Here is a tip: look for a Cask Marque sign (Fig. 5.6) outside the door of the establishment. Under this scheme experts patrol the pubs and ensure that the folks know how to look after the beer and serve it in a way that will delight the customer. Cask Marque also has a website where you can track down recommended establishments (https://cask-marque.co.uk/).

Fig. 5.6. The Cask Marque logo. Thanks to Paul Nunny.

Of course, the long-standing champions of cask ales in the United Kingdom have been the Campaign for Real Ale (CAMRA; www.camra.org.uk). Gratifyingly, these people have eased back a tad on their former insistence that cask ale is the only "real" beer. That was always a mite ridiculous, when one realizes that it would exclude a Belgian Lambic, a Trappist ale, a Czech Pilsner, a West Coast bottled IPA. One needs to remember, too, that beer and pub recommendations from CAMRA likely come from enthusiasts rather than the trained professionals that are employed by Cask Marque. Nevertheless, I rejoice that CAMRA came into being—champions as they have been since 1971 of a time-honored beer style.

Kegged Beer

Cask ale is very much a UK phenomenon. Of course, there is ample draft beer elsewhere in the world, but for the most part this is kegged. The beer may very well have been filtered and perhaps pasteurized and is then filled into kegs of stainless steel or aluminum with a higher level of carbonation (perhaps a little more than 2 mL of CO_2 in every mL of beer) that is usually delivered by adding extra CO_2 to that made in fermentation. The serving temperature for keg beer is usually rather lower than is that for cask ale. In all honesty hand-pumped cask ale is my preference, but it would be silly for anyone to claim that there are not a great many excellent kegged ales and lagers across the globe. View them as just different styles, that is all.

Presentation

Whether it is cask-conditioned ale or kegged beer or indeed anything poured from a bottle or can, the expectation should be that the beer is presented to its best advantage. Compare and contrast these two scenarios.

I was in a restaurant in the Bay Area and a waitress of immense charm came to the table and barked, "What do you want?"

"Well, a menu would be very welcome," I replied.

It came. "What do you want to drink?"

"A Coopers Sparkling Ale, please," I replied.

"Do you want a glass?" she retorted.

"Yes, why do you ask?" I said, reasonably.

"Well, we don't have very many glasses and we don't want to run out."

"Would you have asked me if I want a glass if I was asking for wine?" I inquired.

"Don't be stupid," she said. "Now what do you want to eat?"

Over to Ghent in Belgium and the sheer joy of being presented with an eight-page book filled with a vast selection of beers and, to my immense delight, a listing right at the end under wine that limited itself to "red, white, rosé." Wonderful beers, but every selection presented in exactly the right glass for that beer, with the branding in all its glory. Reverence.

I have long since lost track of the number of times I have had to ask for a glass in the States, only for the beer to arrive in completely the wrong vessel. Is it really so obvious that a lemon-colored lager should not be served in a Guinness glass?

Yes, I know that many an American grew up drinking beer straight out of the bottle and, "Darn it, that's the way I'm going to carry on drinking my god-damned beer." I recall once being in a bar witnessing a seven-foot guy with tattoos down his neck slurping his Bud Light straight out of the bottle and my wife saying to me, "Go tell him." It is okay, I tolerate it. I just wish it were otherwise, that is all. It is just not possible to get a true appreciation of the aroma of a beer if you drink it straight from a bottle or can because the vapors are not hitting your nostrils. Let alone that you cannot see the sheer beauty of the bubbles and the color.

Which brings me to the matter of the shape of the glass. Folks often ask me, "What is the best glass for this beer?" and my reply is, "A clean one that does not have the wrong name on it." A number of years ago I pitched up at the Gordon Biersch bar in San Francisco International Airport. The first thing they did was card me (if you are not American that means asking for proof of age.) Yes, really: upper-end-of-middle-aged bald (with wisps of gray) bespectacled me. I looked at the woman. "Just how badly do you think I have lived?" She replied that they card everybody. I shook my head, sighed and ordered a Gordon Biersch Hefeweizen. Along it came in a Bud Light glass. Nothing wrong with Bud Light, the biggest selling brand in the country. But it ain't Gordon Biersch Hefeweizen.

I would venture to suggest (and there are many who will not be happy to read this) that the shape really has rather little to do with beer quality and is much more a matter of tradition and association with a certain product or style (and that I applaud). It is said that the glass shape can have a profound effect on the foam. I would say that it is far less important than the cleanliness of the glass, the vigor of the pour, and whether there are any sites on the bottom of the glass that promote bubble formation. Others say that the shape influences the journey of the aroma materials into the head, and I say that there is precious little scientific evidence to support that.

I say again, I love the theater of proper beer presentation, and the glass is key to that. Nevertheless, it is primarily a visual and psychological thing.

In passing, I might also repeat my plea not to adorn glasses with decorative items, for example, slices of lemon or lime. One of the more bizarre examples of this came many a long year ago when I hosted a party of visitors from Coors. At lunchtime we all repaired to Café Italia in town. "Six Coors Originals" was the order. Along they came, gratifyingly accompanied by glasses. However in each glass was an olive.

"What's this?" I inquired.

The waiter beamed. "It's a poor man's cocktail," he replied.

"I see," said I and, gesturing to the person to my immediate left, said, "Could you explain that to this man whose name is Peter J. Coors Jr.?"

To return for a moment to the argument often made by American bar owners as to why they don't do the classy thing and present beer in dedicated glasses on a branded basis: the losses would be enormous due to customers stealing the glasses. I heard the ultimate solution to this one time: take a shoe from the customers. When they return the glass, they get to retrieve their shoe.

Clarity

I was always brought up to expect most beers (*Hefeweizen* is a notable exception) to be bright. That is, not cloudy or turbid. At Bass we took enormous pains to ensure that the beer was served with glorious clarity in our pubs. Witness, then, my words about fining earlier. Kegged beers were filtered.

In latter times a storyline has built up around the merits of cloudy beer, originating with the introduction of juicy IPAs (see chapter 3). Fundamentally the storyline pitched by many is that these are superior products *because* of their cloudiness. These people say that to filter beer is to strip much of the stuff that contributes to flavor. The reality is that most of the materials that impact taste and aroma are not removed by filtration. Indeed, turning the argument on its head, many of the materials that are responsible for the cloudiness in these beers, such as yeast or added stuff like starch, have of themselves no impact on flavor.

Of course, it is an argument solely for the purist. These hazy beers are selling rather well, irrespective of what hoary old traditionalists like me say or do. That is the ultimate raison d'être after all: will my beer sell?

Head

Let us turn though to the foam. In my time with the Bass company and later at the university we have done detailed studies to confirm absolutely that people are influenced by the head on beer, and a clear majority of people globally feel that a good head means a better beer.

So the pursuit of a better understanding of the science of foam has been a core part of my research in some 40 years of endeavor. (When I graduated from university with a PhD in biochemistry I thought I would go on to cure cancer. Instead, I tell you how to put a head on beer.) I even established a coalition of like-minded researchers, and we called ourselves the European Brewery Convention Foam Sub-Group. I recall a German saying to me, "So why you? Ze English don't care about foam." I retorted that to visit London, where admittedly the beer can look like cold tea, is not to visit England and if he ventured further north to, say, Sheffield or Newcastle, he would realize just how important the foam is. It has to be a couple of centimeters of stable foam atop a full pint of beer, hence the oversized glasses with a line to mark the pint of liquid and ample room above for the head (Fig. 5.7).

Pretty much anywhere on the globe you go the foam is an indispensable part of the beer-drinking experience. And yet in the States there are people who complain about it and even admit to wiping their fingers on the side of their greasy nose and putting them in the beer to kill the foam. Distressing behavior.

Quite what the good people of the Czech Republic make of this I really wouldn't venture to ask. Their standard pour of beer is *hladinka*, with a glass filled with three-quarters beer and one-quarter foam. But you can get a pour that is essentially all foam. It looks like milk and indeed is called *mlíko*. It's what you might order at lunchtime if you really didn't want to be drinking too much. And then there is *šnyt*, which is one-third beer, one-half foam, and one-sixth empty glass. Again, it's about being seen to have a

Fig. 5.7. An oversized glass. Thanks to the Joseph Holt Brewery.

mug as the same size of everybody else and with the same total volume in it, but with a lesser amount of alcohol. The common denominator is that they all contain foam!

The art of the pour is a thing of beauty. First, the glass must be scrupulously clean. Any grease or fat will bust your head, so to speak. You should never wash beer glasses alongside anything else, such as food plates, whether you are using a sink or a machine.

If we are talking use of a sink, then here is the protocol. Nice hot clean water with a washing-up detergent in it. Beer glasses only (though you can use the water for other stuff later). Wash the glasses thoroughly inside and out. Then rinse away the detergent from inside and out with trickling clean water (because detergent residues will kill foam). Let them drain dry. Do not wipe them on cloths, because there is a chance that there is some grease on that cloth and also a chance that fats from your greasy fingers will make contact with the inside of the glass. You can test whether they are clean by getting the inside wet and sprinkling the glasses with salt. It should stick evenly to the whole of the surface. Any spots where salt is not sticking are greasy. If a beer is poured into such glasses, where those greasy spots are you will see ugly big bubbles at that point (Fig. 5.8).

Next, we need to dispense the beer such that foam gets produced. I remember being in a restaurant in Davis with a senior executive from a major brewing company and we ordered a couple of his beers. The server came along with the two bottles and, thus far to her credit, two glasses. Then she put the tip of her tongue between her teeth and proceeded to pour as slowly and gently as possible the first beer into its glass. Doug was aghast.

Fig. 5.8. Beer in a poorly washed glass. Photo by author.

"What are you doing?'

"I'm trying to stop the bubbles" she replied, innocently.

Doug's jaw dropped.

"Gimme that here," he said and proceeded to show her how it should be done.

Pour with vigor, as I am wont to say. For in so doing you are nucleating the carbon dioxide as bubbles. You cannot get a stable foam if you do not make it in the first place. For small pack it is really quite straightforward: slightly incline the glass but pour vigorously so that the beer splashes into the glass and is essentially completely converted into foam. Obviously, you do not keep on until there is beer cascading out of the glass. Then look at the foam. It starts to change in its appearance from being very wet to being much firmer and almost solid. This is because proteins that originated in the grain are being linked by the bitter substances to form a network that holds the bubbles together.

While this is happening, liquid beer is draining from the foam and space will be created for you to progressively top up (more gently now) from the glass or can and, voila, a thing of beauty. And as you sip it steadily, cathedral windows of foam cling to the side of the glass. My!

(Incidentally, I was in Mount Shasta City one time in a coffee house and headed off to the men's room. I was waiting my turn outside the locked door when a young chap walked by, took a double take, and said to me, "Pour with vigor!" My reply was "What? At my age?")

Beer on tap, too, needs to be poured with vigor and indeed the scenario applies if you find yourself in a Dublin bar asking for Guinness. You must be patient while the glass fills with foam, and then is progressively topped up. In passing I would mention that the foam on draft Guinness (and Guinness in can with the widget, Fig. 5.9) is enhanced by the use of nitrogen gas alongside the carbon dioxide.

Even in countries where they do not take things quite so slowly, there will be time-honored techniques in beer dispense. In Belgium, for example, they will pour with enough vigor to ensure that there is some beer overflowing the glass. They will level the surface of the foam atop the glass with a knife and then stand the glass briefly on a towel to dry the base.

In England and elsewhere there are sparklers on the end of the taps that can be loosed or tightened to impact the force with which the beer is delivered into the glass.

The hand pumps used to dispense cask ale are nothing more than part of a system for pumping the beer up from the cellar. Look next time the barperson

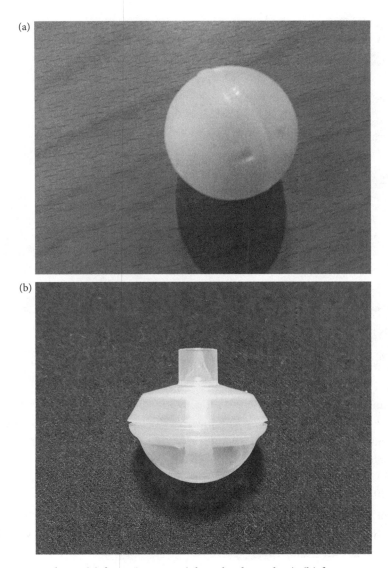

Fig. 5.9. Widgets: (a) from Guinness (photo by the author), (b) from Boddingtons (photo courtesy of Joe Williams).

pulls the pint and see the swirling of the bubbles in the glass and note that he or she will often let the beer stand a short while, allowing the liquid to drain a little from the foam before the pint is topped up.

And then enjoy.

6

Beer Talk

This cuvee flirts with perfection. Already revealing some pink and amber at the edge, the color is surprisingly evolved for a wine from this vintage. However, that's deceptive as the aromatics offer incredible aromas of dried flowers, beef blood, spice, figs, sweet black currants and kirsch, smoked game, lavender, and sweaty but attractive saddle leather-like notes. (Robert Parker, *Wine Advocate* 179, October 2008)

"For heaven's sake" is one's initial take on such affectation. "It's wet and red. Do you like it or don't you? If you do, just shut up and drink it."

I actually feel the same way about music. Consider:

This time it's overloaded with funereal synths and arpeggios that twirl frantically in anguish as if they had nowhere else to go, saturating the cloudy soundscape with particulate matter so intricate it's a wonder all this sound data can be contained in a single mp3, never mind a groove in wax. The fluttering effects are only further confused by the bleary smudge of it all, cinematic and grand but stuck in Burial's world of canned frequencies: The locust-swarm effect of the filters is impossibly stirring, far more visceral than perfect clarity ever could have been. (https://pitchfork.com/reviews/albums/16292-kindred)

Hopefully, for the most part, anyone connected with beer will keep feet firmly on terra firma and in musings about the product be unaffected and straightforward. What exactly does that look like? How can one meaningfully discuss beer?

Let's say we have a bottle before us and a buddy beside. We can start our observations at this point.

"Hmm, you think they would have learned how to get the labels on straight, wouldn't you?"

Not a trivial thing. I well recall as the research manager for Bass having the car loaded with luggage and kids and just about to head off on vacation, only to hear the phone ring and, on picking it up, listening to technical director Gus Guthrie's dulcet Glaswegian tones: "They've just packaged Lamot Reserve for the first time and the labels are skew whiff." This new premium product had to look right from the package and on inward. My wife drove off with the kids and I headed into the brewery. We were both less than pleased.

Back to you and your buddy. What should we be looking for on the label? First, ABV, which stands for "alcohol by volume" and not, contrary to the opinion of one of my students in answering a question, August Busch the Fifth. Just how alcoholic is the brew? We can get a clear idea from this not only of how "heady" the brew is going to be, but also what the order-of-magnitude calorie count is. Alcohol is the main source of calories in beer (just as it is in wines and spirits). And we can also establish how many units of alcohol this represents, a unit representing 10 mL of alcohol. Thus in a 5% ABV beer, one unit is equivalent to 200 mL of that drink. For a 40% ABV whiskey, one unit is represented by 25 mL.

We might also impress our chum by explaining the difference between ABV and ABW, the former being milliliters of ethanol per 100 mL of beer and the latter being grams of ethanol per 100 mL. Now a milliliter of ethanol weighs 0.79 g, so if you had a beer of 1% ABV (which would not be one of my favorites), its ABW would be 0.79. In turn, a 4% ABV beer would be 3.2% ABW. This is the infamous "3-2 beer" that has become the stuff of legends. Lots of people adopt a kind of sneer when they talk about this type of beer, even though it is close to the average alcoholic strength of beer in the United Kingdom and only about 0.5 percent by volume less than the average beer strength in the United States. Of course many people associate it with Utah, mention of which puts me in mind of a visit I made once to that fair state and a trip to a restaurant.

"A beer please," I said, naming a brand (and not Polygamy Porter, famously advertised on a platform of "Why Have Just One?" and "Bring Some Home for the Wives").

The waiter stared at me stonily. "You have to order some food first," which I did. I was thirsty and it wasn't long before I ordered a refill. The waiter looked aghast. "Another?" he said.

"Yes," I barked. "Do I have to order dessert with it?"

Let's get back to the matter of labeling generally. What you are unlikely to find on a beer bottle is nutritional information and a list of ingredients. It is not mandated but, rather, is optional. There are some companies that do offer information on the key ingredients and also nutritional information like protein and carbohydrate content and calories. And, of course, it is mightily helpful to the consumer to be able to read the ABV value, but even this is not uniformly required across the United States. Following Prohibition, Congress decreed in 1935 that labels should not state the strength of beer, arguing that if offered two beers at the same price, consumers would opt for the stronger one. If they didn't know the strength, then they would just as likely buy the weaker one. (Which seems bizarre to me, because surely a little bit of experimentation would soon tell you which was the bigger bang for the buck.) It is actually a state issue, so in New York no ABV information is allowed, whereas in North Carolina, Washington, and New Hampshire there has to be an ABV declaration for beers containing more than 6%, 8%, or 12% respectively.

The ABV of many beers is listed on the menu board in craft breweries. However, many in the smaller companies worry about legislation coming in to insist on the declaration of ABV and other ingredients for the simple reason that they don't have the analytical capabilities and perhaps control strategies to be able to reliably deliver a product of a given composition consistently.

One piece of information that any brewer could offer and in my opinion should offer that gives hints to the age of the product—a packaged-on date or at least a best-before date, as I discuss in chapter 4.

Moving from the label to the glass bottle itself, we can draw attention to the pristine nature of the container. In every likelihood in the United States the bottle will be undamaged. However, in Canada it is probable that there will a couple of bold white scratches or rings in the lower half of the glass (Fig. 6.1). These are scuff marks and are caused by the bottles repeatedly passing through conveyors in the packaging hall. You see, in Canada (and many other countries) they use returnable glass. Empty bottles are returned, sorted, washed, and refilled. In this journey there is plenty of opportunity for the bottles to rub against one another and against the sides of the conveyor belts. Some would have it that it must be more environmentally responsible to reuse glass, as opposed to the employment of virgin glass every time. But bear in mind the environmental costs of collecting glass, of sorting it, and the intense cleaning and sterilization regimes that need to be employed to get the containers ready to be refilled. Oh, and next time you casually stuff an empty chip packet (or

Fig. 6.1. Scuffing. The middle bottle displays typical scuffing in returnable glass. Photo courtesy of Alex Speers.

worse) into such a bottle, realize that somebody in the brewery has got to get it out.

You might mention at this point plastic bottles and cans, and how beer is able to be kept fresher in a can than in a glass bottle because air that promoted staling can't get into a sealed can but it can get between the neck of the bottle and the crown cork. Worst of all is a plastic bottle, because not only can air get in through the closure but also through the walls of the container. You might draw attention to the color of the glass, saying that light can get most readily through clear glass, then green glass, but much less easily through brown glass and that light reacts to cause the bittering materials from the hops to break down to give the whiff of skunks.

By now we are thirsty, so it is time to pour out the beer. First off, we need to open it. It is possible that the top is a twist off. Be careful, it may well not be, requiring a "church key" for reasons discussed in chapter 4. Embarrassing if you tear skin trying to wrench off a pry-off crown cork.

By this time you might also have had another awkward moment. I certainly had this the time I was in a major beer company executive's home.

I was there to beg for a major donation (of the order of $5 million) and was standing across a kitchen counter from him, with a bunch of bottles of the legendary beer between us in an ice bucket. He handed me one and I clutched it. Of a sudden I felt movement between my thumb and fingers. The label was sliding off the bottle, which receptacle was slightly tilted away from me and unfortunately positioned such that when the label totally came away, remaining in my clammy hand, the full bottle of beer was now rocketing in the direction of my host. We laugh about it now. But the moral is that the brewer needs to choose glues that ensure good contact with the bottle, no matter how damp the product gets. Oh . . . and pour the beer as soon as possible (though properly) into a glass.

Okay the label has not slid off. We get to open the beer and all of a sudden whoosh! The beer has exploded out of the container spontaneously and soaked everything: the table, the floor, and the pants. You explain,

> That, I am afraid, is gushing. The carbon dioxide does not normally get converted into bubbles that readily. That's why you have to put some effort in to produce a good head on beer. But sometimes there is a gushing problem. One cause is that a package has been shaken or dropped—and I think I can trust you enough to be confident that you weren't playing a trick on me. So that means that there is a gushing promoter in the beer. These are very small particles (usually you can't see them) that may be of various types. By far and away the main cause is a small protein that comes from a mold called Fusarium that can populate grain when it is grown in wet conditions. It's seldom a problem in North America, but can be a big deal in Northern Europe.

Usually, though, this gushing does not happen. We get to pour it out normally and we are going to explain that this is not only so that we can get a full appreciation of the beauty of the product, but also because otherwise we don't get a full appreciation of the flavor. Most of this is detected by the nose and not by taste. The main flavors sensed by the tongue are sweet, salt, sour, and bitter. Most of the character, though, is assessed by smelling. Prove the truth of this by asking your pal to pinch tight her nose and taste the beer. Then suggest she free her nostrils and take another sip. There will be a "wow" utterance right now. The thing is that if you drink the beer straight from a bottle (or can), you do not get the aromas surging into your nostrils, whereas you do if you pour the drink into a glass and dangle your nose in the headspace.

So we pour. And how do we do it? Pour with vigor, as we saw in chapter 5. How do we articulate the nature of the foam? If we can leave the bottled product for a moment and shift to draft beer, we might comment on its depth, of course, passing opinion on whether we feel shortchanged if the relative proportions of foam and liquid beer in the glass means that we are certainly not getting a full pint. Such may very likely be the case in many a European country, but certainly not in the United Kingdom, where questions were asked in Parliament and legislation established that an imperial pint of beer does not include the foam, hence those oversized glasses in the more northerly reaches.

We might offer a judgment on the color of that foam. Usually it is white but sometimes it may well be tinged amber or brown, a reflection of some of the sizable color-forming molecules being carried into the head. If you look carefully at a pint of Guinness you will see that despite the superb blackness of the beverage the foam itself is dazzlingly white, and close inspection will make you realize that this is a function of the bubbles being extremely small. Small bubbles make for much more stability and, in the case of this product and others, is impacted by the use of nitrogen gas alongside the carbon dioxide. The nitrogen also leads to a smooth mouthfeel to the product, which is only in part due to the fact that the rich small-bubbled foam gives it a luscious texture.

Having admired the beauty of the foam, we now might comment on its longevity and how well it is retained as we progressively sip the liquid. And of course we can explain to our pal that the beautiful cathedral windows of laced foam that cling to the side of the glass as we quaff are a result of the proteins from the grain being joined together by the molecules from the hops that give the beer its bitterness and that the resultant almost solid nature of the foam is what makes it adhere to the walls.

Of course the foam is not the only visual aspect of beer to be commented upon. There is of course the clarity. You might have a *Hefeweizen* before you and, having pointed out that this style tends to be the most highly carbonated of all the beers, you would also draw attention to the fact that the cloudiness is due to the presence of yeast (*Hefe*) in the product, which in turn is made from wheat (*weizen*, or white, *weissen*). You may mention that customarily in Germany the affectation of a slice of lemon is not the done thing—though you may draw a line on referring to the former head brewer of a small facility in California who referred to "NFL" (no f***ing lemon). I recall giving a talk to UC Davis alumni in Portland, Oregon, coming out with one of my very regular lines that there are three indicators of an authentic *Hefeweizen*: the smell of cloves, the smell of banana, and no lemon. When the statement had already left the departure lounge of my brain and was making its way to the runway for flight from my mouth, I spotted the scion of a famous brewing company whose *Hefeweizen* is well known for no clove, no banana, but a slice of lemon. Soon after concluding my remarks I went up to him to apologize. "It's okay," he said, "I have sold a lot of it."

In current times chances are that your beer is as likely cloudy as not. You might comment on the fact that it is actually easier to make a beer that is not cloudy than one that is, because it is easier to take away the crud (using filters, finings, or even centrifuges) than it is to make sure that it is there at the same level every time. You might mention that there is very little evidence that the clarification of most beers has any negative impact on the flavor and that, at the other extreme, there are some brewers who claim that their highly murky brews taste great because of the cloud but who are adding it in by using flavorless starch or other materials.

Abject turbidity can make it sometimes difficult to truly admire the color. Impress your pal by telling him that there are beers of every hue from colorless to black. Colorless beer is not achievable absent some way of stripping the color out of the product by some form of filtration, notably practiced a few years ago by the Miller Brewing Company with Miller Clear Beer, which tanked after a very few weeks. You can explain that the color primarily originates from materials that are produced in the malthouse during the kilning process and any subsequent roasting in drums. Yes, you might point out, there has been something of a history of using caramel to add color and flavor to beer, but it is relatively rare nowadays. You may draw attention to the perils of oxygen creeping into the brewing process and also the finished beer in that one consequence is the oxidation of polyphenols that come from the malt and from the hops, this oxidation turning them reddish-brown. This is the selfsame thing that happens in a sliced apple. And of course you may care to give examples of beers that have colors that deviate from the mainstream, notable examples being *Berliner weisse* to which woodruff (*waldmeister*) essence has been added and also the stuff that the Americans use in their St. Patrick's Day obsessiveness to turn everything, including the beer, emerald.

At last we are ready to talk flavor. And here at last we have some words readily made available to help us. It was in 1979 that Morten Meilgaard of the Stroh Brewing Company in Detroit and Charles Dalgliesh and John Clapperton of the Brewing Research Foundation in Surrey, England, came up with the flavor wheel (Fig. 6.2). In truth it was originally designed to facilitate brewers talking to one another about what they were tasting and smelling in the beer. The words were chosen for their in-house utility rather than as appealing words that might be used for marketing of beers. Hence the word odor, rather than aroma, at the heart of the wheel. Odor makes me think of armpits. No way would a wine person use that term.

Actually the words around the fringe of the wheel are a latter-day addition and for the most part are more helpful in articulating the nuances of the flavor of any given beer.

More recently in a project led by a former student of mine, Lindsay Barr, a newer concept, the Flavor Map, has been developed (Fig. 6.3).

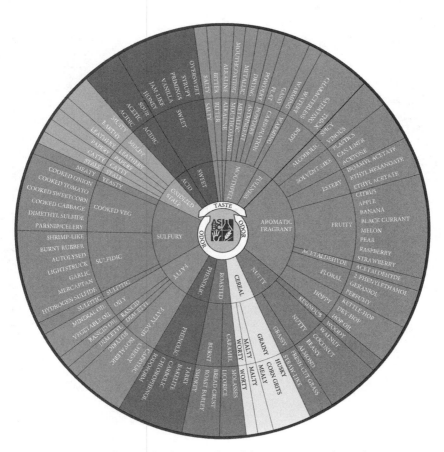

Fig. 6.2. The beer flavor wheel. Reproduced, by permission, from the American Society of Brewing Chemists—© 2009.

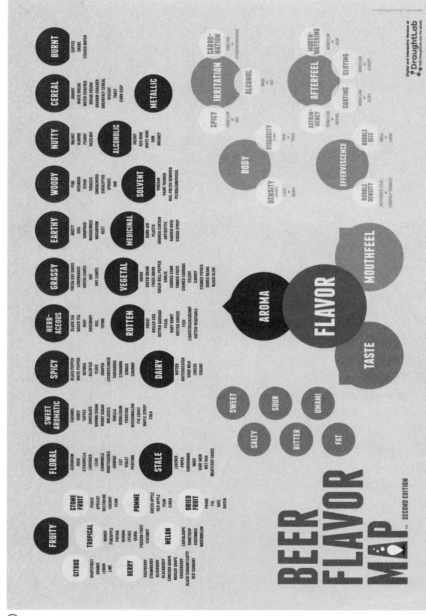

Fig. 6.3. (a) Beer Flavor Map, (b) Malt Flavor Map, (c) Specialty Malt Flavor Map, (d) the Hop Flavor Map. All maps courtesy of DraughtLab (www.draughtlab.com).

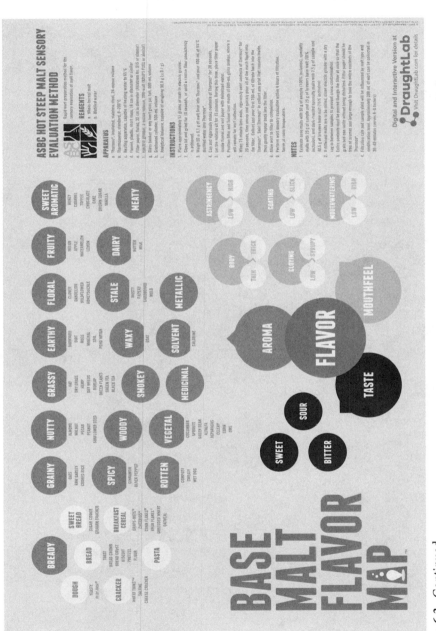

Fig. 6.3. Continued

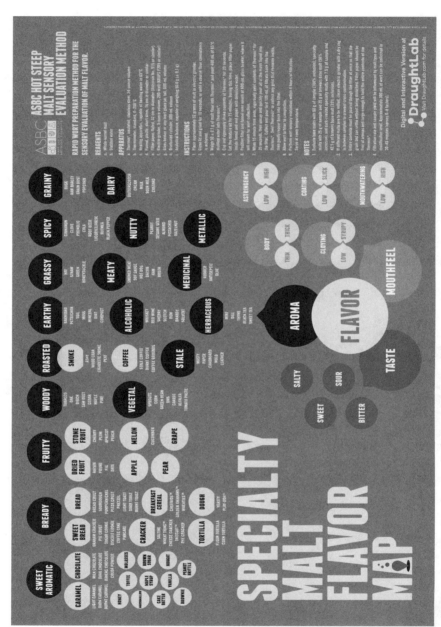

Fig. 6.3. Continued

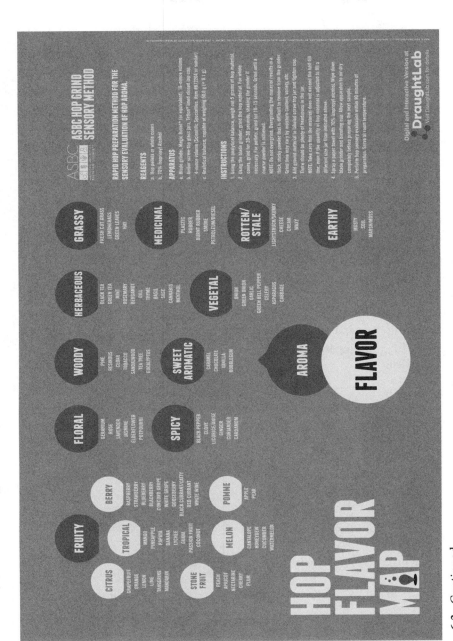

Fig. 6.3. Continued

It should be clear to the reader, then, in these days of extreme, if not bizarre or even ludicrous, concoctions incorporating some outrageous materials, it is a rather more complex scenario than was the case hitherto. Let us just restrict ourselves to some of the more long-standing ingredients that one can link through to the aromas and tastes shown on these diagrams.

As we saw in chapter 2, malt is the soul of beer. The bedrock of any beer is a relatively pale malt that is not kilned so intensely as to destroy the enzymes that are needed to convert starch into fermentable sugars. Such pale malts therefore can have reasonably delicate flavors. The more intense the heating, the more flavor and color is developed. Table 6.1 presents a simple breakdown of the flavors from different types of malt. And so a Pilsner, for example, may very likely be produced with 100% Pilsner malt. A pale ale may be something like 90% pale ale malt and 10% crystal or caramel malt. A brown ale will have a still greater proportion of caramelized malt, while a porter will match its preponderance of pale ale malt with significant amounts of malts like chocolate malt. A stout is more likely to balance its bulk of pale ale malt with chocolate and black malt and often roasted barley. Take a look at chapter 3 to get a better feel for the types of malts likely to be used in given beer styles.

Hops are the spice of beer. To find words with which to articulate their glories in imparting bitterness and aroma, go to chapter 2.

Table 6.1 The flavor delivery from a selection of malts

Malt	Color	Flavor
Pilsner	Pale yellow	Gentle "malty" where the malty flavor is what you would get from eating malted milk balls or cornflakes; slight honey
Pale Ale	Light amber	Stronger "malty"
Munich	Gold	Sweet, "malty," slight fruity
Pale crystal	Amber	Mild caramel
Crystal	Dark amber	Caramel
Dark crystal	Very dark amber; red	Burned fruit
Amber	Amber	Dry biscuit, baked
Brown	Brown	Smoky, spicy
Chocolate	Dark brown	Bitter, harsh, roasted
Black	Black	Same notes but more intense
Roasted barley	Dark brown, black, reddish hue	Burned, smoky, acrid

Yeast can deliver a host of different aromas to beer. The majority of strains of *Saccharomyces cerevisiae* and *Saccharomyces pastorianus* produce the same substances, but maybe in different balances. Thus they all produce so-called esters, which give fruity aromas like banana, apple, and pear. They all produce substances that contain the sulfur atom and that give more vegetable aromas like canned corn, cabbage, and potato. They all produce so-called short-chain fatty acids, imparting aromas like wet dog and goaty. The list goes on, but we can say that this balance of products is not only a function of the yeast strain, but is also influenced by the conditions in which the yeast finds itself: the strength of the wort (Plato value), the temperature, how much oxygen it encounters, and more besides.

It is also undoubtedly the case that these aroma contributions from the yeast can be hidden if the beer contains a "big" malt or hop character. I well recall taking a party of academics who have focused their lives researching the intricacies of brewing yeast to a famous brewery by the coast in California. They were welcomed by the legendary owner, who said that his iconic beer was largely a function of the malt and the hops and that as far as he was concerned the yeast was just a convenient way of making alcohol and carbon dioxide.

Perhaps he overstated the point a tad. And it is certainly the case that there are some yeasts that do produce flavors that other yeasts don't. A good example is the variety of ale yeast that is used to make *Hefeweizen*, which is a brewing strain that is capable of making that clove-like aroma. People also talk about so-called Belgian funkiness in reference to some spicy notes that can be associated with some of the beers from that country.

The ultimate in funkiness comes from a completely different yeast, *Brettanomyces*. This strain was named in honor of the British in the late 19th century, who insisted that it was necessary to condition their ales. It delivers barnyard character, variously described as wet horse blanket and mouse pee. It is most certainly one of the numerous organisms that are involved in the production of wild beers of the Lambic type. Such beers also, of course, depend on the presence of acid-forming bacteria, notably lactic acid bacteria. Wild yeasts and beers are the enemy of the majority of beers. But not so sour beers.

Some of these sour beers are also produced with the aid of fruit, as we saw in chapter 3 when discussing the likes of *kriek* and *framboise* and the Belgian *wit* beers. The latter are of course good examples of the historic use of spices in beer, in the instance of *wit* it being coriander.

Our celebratory discourse on the flavor of beer should also involve a discussion of what we might call "linger." Let me give an example using another
foodstuff. Being from the north of England and possessed of more DNA of
Scandinavian extraction than any other genetic suit, I love kippers, which
are smoked herrings. They make for a wonderful breakfast, accompanied by
lashing of richly buttered bread. About an hour after eating them you fondly
think, "My, those kippers were good." At lunchtime you are thinking, "Hmm,
I can still taste those kippers." After work you head home and kiss your
partner, who says, "Have you been eating kippers?"

Now, of course, I do rather stretch the story out a tad and, yes, I am well
versed in the use of toothpaste and breath freshener. But you get the point.

Some flavors linger. Some don't. There's many a brewer who is firmly of the
opinion that what you should be looking for is a "crisp finish." And who in-
variably snaps thumb and middle finger when saying those words. What this
brewer is referring to is the phenomenon whereby you taste a beer; its flavor
registers with you; you say, "That's nice"; and then the flavor vanishes. Your
palate is nice and clean and ready for the experience to be repeated. Again.
And again. Until the glass is emptied. And until you ask for another.

Some people believe this is the recipe for drinkability, moreishness,
sessionability, whatever you want to call it. I am not so sure. I certainly used
to believe this to be the case. It remains a fact that I don't stick at just one
when I am having a black IPA.

And I still love my kippers.

7

Dining

We were in the Big Room at the sumptuous Sierra Nevada brewery in Chico (Fig. 7.1). The occasion was a multi-course dinner, all prepared by the company's fabulous chef and his outstanding team. The concept was that each course would be paired with a beer and with a wine and that the diners would get to vote whether the greatness of the grain or the wonder of the wine made for a better match, dish by dish (Figs. 7.2, 7.3). I was to emcee the event and Ken Grossman's brother, Steve, spoke to the beers. The wine company sent a brace of boosters, a sales guy and a technical guru, to double-head it on the wine front.

Ken gave me strict instructions (albeit with a glint in his eye) to behave myself. Which I did, though clearly the oenophile duo were detecting a certain teasing tone in my delivery.

The first round was a no-brainer. The beer won hands down. Just as I was about to head to the microphone to start round 2, the wine technician darted to the stage.

"Well, there is going to be no contest here. This wine speaks pork," and he proceeded on a five-minute tirade about the wonders of the beverage to hand and how it was a majestic marry for anything that emerged from a pig.

Ken looked at me. "You or me?" he asked. I said I would do it and I took to the podium.

"If you look at the menu, you will find that the pork is actually duck. Pork is the next course."

The house erupted and a white serviette was fluttered by the wine marketer.

"We give up. This is home-field advantage."

Actually it was no such thing. It was simply that beer is usually a far better accompaniment to food than is wine. There should be no surprise in this. Think for a moment about the diversity of foods on God's earth. Virtually limitless. Think now of the breadth of beers that are available, from the palest and blandest to the biggest and boldest, beaming out as they do characteristics derived from grain, water, hops, yeast, and diverse other components. It is quite evident that there is a huge diversity, demonstrably greater than the

Fig. 7.1. The author opening the beer and wine dinner in the Big Room at Sierra Nevada in Chico.

breadth of flavor attributes available from wines. It is only too obvious that there is greater scope to find matches between food and beer than between food and wine.

That extends to any foodstuff, even to some of the less common eats in the West. Thus was I persuaded to do a Bugs and Beer tasting once, with me matching the beers to a selection of insect-derived dishes (Fig. 7.4), chosen and presented by the intriguing David George Gordon (Fig. 7.5). A couple of days before the event the organizer Elizabeth Luu (Fig. 7.6) and I went into the studios of our local station, Capital Public Radio, to promote the occasion on the excellent *Insight* program hosted by Beth Ruyak. The chef was on the line from Seattle. We presented Beth with the dried mealworms and I swear she was close to retching. "That's all we need," said Mr. Gordon, "a fussy eater."

I am not so jaundiced as to say that there aren't some pretty decent pairings to be had with wine. A Syrah is rather sublime alongside some roast beef of old England. And a Sauvignon Blanc sure marries well with poached trout. But I would venture that for each of these dishes a ruby red ale and perhaps a *Gose* respectively work even better, quite apart from the added boost to the bolus lubrication to be had from the increased proportion of H_2O.

WELCOME (as guests arrive):

Pale Ale & 2010 Chardonnay

COURSE 1

Local fresh water prawn tamale, toasted spent grain and corn masa tamale filled with Cascade hop steamed prawns. Served with Hellraiser pumpkin mole sauce and Viognier red chili jus.

Pair with: Hellraiser & Viognier

COURSE 2

Masa Farms organic wheat flour and Malted barley Pappardelle pasta. Duck legs marinated in local preserved oranges and Celebration Ale, porcini mushrooms, roasted grapes and fried parsley.

Pair with: Celebration Ale & 2008 Syrah

COURSE 3

Porcetta, organic pork tenderloin brushed with porter mustard and rosemary, stuffed with Fra Mani mortadella and salami rosa. Ovila Quad marinated bacon fat fried spent grain brioche bread, pork cheek broth spiked with stout mustard. White, pink and purple carrots glazed with Pale Ale malt, mountain spinach sprinkled with warm ver jus (from SN garden).

Pair with: Ovila Quad & 2008 Merlot

COURSE 4

Estate filet mignon. Served with Porter demi glaze and wine demi glaze, truffle oil roasted Estate potatoes, Estate turnips and carrots.

Pair with: Porter & 2008 Cabernet Sauvignon

DESSERT

Sampler array: Spent grain brioche filled with caramelized local Fuji apple charlotte. Bourbon Barrel aged Trippel donut with Ovila Saison glaze. Honey persimmon mousse, malted spice cookie.

Pair with: Bourbon Barrel-aged Belgian Trippel & N.V. Mistelle de Viognier

Fig. 7.2. The menu for the beer and wine dinner. Image courtesy of Sierra Nevada Brewing Company.

Fig. 7.3. Two of the courses: (a) prawn tamale; (b) porcetta.

I think the best way to illustrate the argument is if we turn to cheese. Worldwide there is the notion of the cheese and wine tasting. But as someone who has led a huge number of beer and cheese events alongside cheese experts, let me tell you that absolutely the possibilities are vastly greater with beer.

Let's home in on one specific example. Humboldt Fog and Beatification from Russian River. A goaty richness in the cheese sliced through by the acidic funk of the beer.

Here is a marvelous example of one of two principal ways in which a pairing should be based: one characteristic *balancing* another.

The other way in which we can seek to establish a match made in heaven is by choosing a beverage that *reinforces* the flavor of the food. In the case of beer a good example would be a *Rauchbier* accompanying a smoky cheese, such as a smoked Gruyère, or some pork smothered in barbeque sauce.

In all my years of leading beer dinners I learned that the later in the evening you get, the better the pairings become. Now there is one very obvious reason why this might be. However I maintain that when it comes to the cheese course but also the dessert course there are some marvelous possibilities. Without doubt the greatest triumph I ever delivered was a dark chocolate gateaux paired with a Narwhal Imperial Stout aged in bourbon barrels.

Fig. 7.4. Some of the courses.

RMI
Robert Mondavi Institute
for Wine and Food Science
UCDAVIS

Saturday, November 1, 2014: Bugs & Beer Menu

1. Teriyaki Grasshopper Kabobs (flavor = teriyaki seasoning, plus like green peppers; some people say it tastes like chicken)
   - Pairing: Rubicon Angus Scottish Ale
2. Dried Mealworms (flavor = roasted nuts/seeds, mushrooms)
   - Pairing: Ruhstaller Gilt Edge Lager
3. Sago Worms with Wasabi Dipping Sauce (flavor = wasabi flavored pistachios)
   - Pairing: Lagunitas Pils
4. Baked European House Crickets (flavor = nutty shrimp, cajun seasoning blend)
   - Pairing: Sudwerk Hefeweizen
5. Cambodian Crickets (flavor = earthy, somewhat bitter taste)
   - Pairing: Gordon Biersch Winterbock
6. Ant and Pear Salad
   - Pairing: Sierra Nevada Boomerang IPA
7. Chocolate-dipped Chapulines (salt and vinegar potato chips with chocolate)
   - Pairing: Berryessa Whippersnapper English Mild
8. Cricket Flour Cookies
   - Pairing: Heretic Chocolate Hazelnut Porter

Fig. 7.5. The menu for the beer and bugs dinner at the Robert Mondavi Institute.

The look on the punters' faces said it all: "Shut up, Bamforth, and let us enjoy this moment in heaven in peace."

I was at the Sheraton San Diego Hotel and Marina's Harbor's Edge restaurant. The seafood linguine was spectacular. And my first sips of Dock of the Bay IPA said to me, "Good choice." It remained a good choice, but there was a distinct change in flavor as I progressed, such that as I ate more of the fish dish, so did the hoppiness of the ale decline as its toffee/caramel character increased. This matter of food and drink pairing is a movable feast to

Fig. 7.6. Bamforth, Luu, and Gordon. You can read more about Mr. Gordon at http://davidgeorgegordon.com.

the extent that changes are happening as you proceed through the dining experience.

Another truism is this: the food should not overwhelm the beer and vice versa. I adore Indian food. No disrespect to a Bud Light, but it just won't hack it when confronted with lamb phall. Equally you wouldn't want to quaff a Sierra Nevada Hop Hunter if you were nibbling on a cucumber sandwich sans crusts.

8

Body

It is rare indeed that my audience is entirely female. In fact, I can recall just two occasions. One was a meeting of the Pink Boots Society, the organization for brewsters. The other was a gathering of dieticians.

This latter event was in Walnut Creek, California. I was invited to talk about beer and health. There was only one issue that I uttered in expectation of raised eyebrows. I told them that, back in England in 1980, when my wife gave birth to our first child, there was beer to be had in the nursing ward. I always say that this was ostensibly to promote lactation but was, in reality, probably more of a reward for a somewhat strenuous experience. I noticed a few nods in the audience, but afterward several of these women came up to me and said, oh my word yes, they had ample experience of young and older moms alike insistent that a beer a day was fantastic for successful breast-feeding.

Can you imagine if I wandered today into a Sutter or Kaiser hospital with a couple of cases of beer and declared that it was for the new moms on the nursing ward? I would be rapidly turned away. It is a very different world nowadays. Look at a beer label in the United States to read these words:

(1) According to the Surgeon General, women should not drink alcoholic beverages during pregnancy because of the risk of birth defects.
(2) Consumption of alcoholic beverages impairs your ability to drive a car or operate machinery, and may cause health problems.

Okay, they are not saying anything here about having a beer *after* pregnancy. And for sure no sensible person advocates drinking to excess on any occasion, least of all if you are supposed to be in control of a potentially lethal device. However the general message being touted in latter times is that there really is no safe level of drinking—for anyone.

In the September 22, 2018, issue of *The Lancet*, Professor Emmanuela Gakidou of the University of Washington published a 20-page paper of which seven of the pages were devoted to listing the authors contributing to it, entitled "Alcohol Use and Burden for 195 Countries and Territories, 1990–2016: A Systematic Analysis for the Global Burden of Disease Study 2016." To state verbatim their conclusion:

> Alcohol use is a leading risk factor for global disease burden and causes substantial health loss. We found that the risk of all-cause mortality, and of cancers specifically, rises with increasing levels of consumption, and the level of consumption that minimizes health loss is zero. These results suggest that alcohol control policies might need to be revised worldwide, refocusing on efforts to lower overall population-level consumption.

Scary stuff. Until, that is, we do our best to delve into what the authors are saying. In itself, that is not easy, the article being written as is normally the case for publications in the medical and scientific worlds not for the interested layperson but rather for others in the fraternity/sorority.

What we have is fundamentally an exercise in statistical correlation. On the one hand there have been laudable attempts to get a handle on how much individuals are drinking in countries across the world. On the other hand there is a compilation of the diverse diseases and other problems, such as transport injuries, self-harm, and so on. And then there is a drawing of correlations between the two sets of data. Conclusion: increased alcohol consumption correlates with an increased risk of problems.

I will not make the cheap throwaway comment that it would be very easy to conclude that the drinking of water each day is associated with some terrible repercussions if we unquestioningly try to draw correlations of this type. Probably 100% of people dying from whatever cause probably drank some water in the previous 24 hours, so, interpreted blindly and unthinkingly, there would be a scary correlation to be drawn. Silly of course, but you get the drift. Rather I will turn to an interpretation of the study by Professor Sir David Spiegelhalter, chair of the Winton Centre for Risk and Evidence Communication at the University of Cambridge. This learned statistician criticizes the Gakidou study for not reporting *absolute* risks; rather it gives *relative* risks. Spiegelhalter applauds *The Lancet* for responding to this criticism and responding as follows in a press release:

> Specifically, comparing no drinks with one drink a day the risk of developing one of the 23 alcohol-related health problems was 0.5% higher—meaning 914 in 100,000 15–95 year olds would develop a condition in one year if

they did not drink, but 918 people in 100,000 who drank one alcoholic drink a day would develop an alcohol-related health problem in a year.

This increased to 7% in people who drank two drinks a day (for one year, 977 people in 100,000 who drank two alcoholic drinks a day would develop an alcohol-related health problem) and 37% in people who drank five drinks every day (for one year, 1252 people in 100,000 who drank five alcoholic drinks a day would develop an alcohol-related health problem).

Let us just look at the one-drink-a-day scenario. What is being said is that if 100,000 people abstain from alcohol, 914 of them would develop one of the health problems referred to in the study. If another 100,000 all had one drink per day (a drink being defined as 10 g of alcohol), then 918 would get a problem. Seems like a chance worth taking to me. Even at two drinks a day it becomes an extra 63 people out of the 100,000 that would suffer. At five drinks per day 334 out of the 100,000 would be in the problem zone (or 0.3%).

Spiegelhalter spells it out in another way. He says that if 25,000 people between them consumed 400,000 bottles of gin (750 mL), just one of those people would develop a health problem as a consequence. For a rate of alcohol consumption a tad over that recommended as a maximum in the United Kingdom (14 units per week) the equivalent numbers would be 50,000 bottles of gin (or more than 900,000 bottle of beer) divided between 1,600 consumers, leading to just one of them developing a problem.

The Cambridge statistician took issue with the Seattle authors' proposal that public health bodies "consider recommendations for abstention." Spiegelhalter points out that there is no safe level of driving either, but governments do not recommend people not get behind the wheel (provided they are sober, of course).

"Come to think of it," says Spiegelhalter, "there is no safe level of living, but nobody would recommend abstention."

More realistically, perhaps, the good professor said that drinking the recommended maximum amount of alcohol is less dangerous than "an hour of television watching each day, or a bacon sandwich a couple of times a week."

None of this represents an argument for ignoring the undesirability of excessive alcohol consumption. Go into any sizable city in the United Kingdom on a Friday and Saturday evening to witness reprehensible displays of booze-fueled outlandish and antisocial behavior. I can only assume that this is what was in the mindset of Professor Dame Sally Davies, chief medical officer for

England, when she presided over the reduction in the recommended alcohol limit for men to match that for women, namely 14 units per week. At a stroke she pooh-poohed diverse studies (that we will come to momentarily) showing a link between moderate alcohol consumption and a reduced risk of cardiovascular problems as "an old wives' tale." (Best not to mention that the Gakidou study **did** support the contention that alcohol can help in this matter.)

Surely it is a matter of education and not legislation? Give consumers balanced and properly reasoned information on which to make their selections. And don't try to penalize moderate consumers who enjoy their tipple by ramping up the "sin tax" as a mechanism to attempt to lessen the debauchery of those who do not treat alcohol as a pleasurable and maybe beneficial component of a well-balanced lifestyle but rather as an adjunct to excessive displays of hedonistic outrageousness.

There are many neo-prohibitionists out there. They might like to consider some of the outcomes of the United States' ill-judged Prohibition experiment from 1920 to 1933. There was a serious downturn in the performance of companies in the fields of amusement and entertainment, including restaurants, which struggled by not being able to serve alcohol. A great many jobs were lost, not only those in the breweries but also people employed all the way from raw material and packaging supplies through to those at the retail end. Governments suffered at all levels—after all, they have known globally for eons that tax on alcohol is a great source of revenue. There were bizarre outcomes. As pharmacists could legally dispense whiskey for a diversity of ailments, there was a remarkable growth in the number of pharmacists across the States. Because wine in church was legal, there was a fascinating increase in the numbers of folk who had appeared to find God. Illicit alcohol served in speakeasies was of dubious quality, and the very existence of such places depended on a disturbing amount of corruption in law enforcement. And perhaps the bottom line was that people were now actually drinking more than they before the Volstead Act.

Empty calories?

A notable university teacher of nutrition hereabouts would speak of beer being "empty calories." I rather think the intent, analogous to the preachings of the aforementioned chief medical officer, was to frighten students into eschewing the noble beverage. If ever there was fake news, this is it. The simple reality is vastly different. Beer contains a diversity of components that are very much capable of being a valuable contributor to the diet when the beverage is consumed in moderation.

Let us begin with silicon. This is located (in the form of silicate) in the husk of the malt and, during brewing, is swept into the beer. The levels to be found in beer were shown by Troy Casey and myself to average 16 mg per liter of alcohol-free beer and up to 41 mg per liter in IPAs. Now there are no guidelines that I can find in any country's documentation that indicate the recommended level for daily silicon consumption, but it is known that silicon deficiency results in bone deficiencies (osteoporosis), reduced levels of cartilage and collagen, and poorly formed joints. So assuming that a healthy intake of silicon is desirable, we might compare the levels in beer with those in other foodstuffs. A half cup (50 g) serving of granola (another good source) would deliver about 6 mg of silica. A similar helping of oat bran would deliver 12 mg. A 12-ounce IPA would offer 14.5 mg. I guess I know which way I would prefer to take my silicon. And, no, I am not advocating moistening your breakfast cereal with beer.

Okay, not convinced? Then let's turn to vitamins. There have been plenty of reports of the vitamin content of beers over the years. Obviously, beer being aqueous, the levels of fat-soluble vitamins are vanishingly small—but not so some of the B vitamins. True, thiamine (vitamin B1) is always deficient in beer. However, others in the cluster can be in significant quantities. Let's just stick with one, folate, which is needed for all manner of cellular functions. Janel Owens, Andrew Clifford, and I actually reported less folate in beer than some folks had claimed, being very critical of the methods employed by others, but nonetheless we still found that beer could supply up to 6% of the recommended daily allowance for this vitamin, about the same that you would get from orange juice. This by the FDA guidelines is certainly not a "good source" of this vitamin, as such foods must deliver 10%–19% of the recommended daily intake. It certainly isn't a "high source," meaning 20% or more of the suggested daily consumption rate. It is however not zero, unlike, say, red wine. Just saying.

When it comes to red wine, the drink most especially touted for his healthful properties, there is a big song-and-dance about antioxidants. This is problematic.

When people talk about this food and that food, this drink and that drink, as being a good source of antioxidants, what they are talking about is a comparison of them on the basis of tests that are conducted in the laboratory. Mostly they use an assay featuring something called Trolox as a standard screen for total antioxidant content. If a food or beverage scores high, then it is said to be a rich source of antioxidants, and all manner of claims are made on the label. Conversely, if it performs poorly, then it is said not to be a promising supplier of these molecules.

The reality is that different laboratory tests do tend to rank foods and drinks in different orders. A product may fare well with one test but not so well with another. However, my main criticism is that just because something shapes up well in a test tube, that is a far cry from what is expected of it in the diet. An antioxidant must get into the digestive system and be delivered to each and every part of the body, where it is expected to protect against the damaging effects of reactive forms of oxygen (hence the name, antioxidant). The number of studies that have given an indication that the antioxidants in various foodstuffs actually do get to the target sites are few and far between.

One of the few that stand up to scrutiny was conducted by Bourne and colleagues from Guy's Hospital and King's College London and BRF International in 2000. They compared the uptake by the body of an antioxidant called ferulic acid (which is also the precursor of the clove-like flavor in *Hefeweizen*). They compared uptake from a tomato and from alcohol-free beer and quantified assimilation in terms of the amount of ferulic acid peed out by the subjects. The more ferulic acid in the urine, the more readily it has been absorbed and the surplus excreted out through the kidneys. There was more ferulic acid in the pee when the consumers had taken beer than when they consumed tomatoes, the conclusion being that not only was beer a good source of this antioxidant, it actually is a better way to get it into your system than when it is in the form of a solid foodstuff, albeit one that can be somewhat mushy.

I am unaware of comparable success stories with the antioxidants from, say, red wine, which tend to be much bigger molecules that are likely less readily transported into and through the system.

Let's now consider fiber. Beer is loaded with it. Okay, it gives you gas, but at least you know you're alive. And, absent discretion, so might everybody else. Seriously, though, beer typically contains about 2 g of fiber in every liter, which originates in the grains used to brew the beer. The Daily Value (DV) guideline for fiber is 25 g per day. Which means that a liter of beer constitutes a good source of fiber.

Not only that, recent research in our laboratories has confirmed that some of the fibrous materials from the grains are broken down to much smaller molecules that find their way to the most southerly end of your digestive system, there to support the beneficial colonic bacteria. In other words, they are prebiotics. In this instance they are not in vast quantities, but present they are.

In summary, then, beer is far from being "empty calories." It does not, of course, constitute of itself a well-balanced diet. Let nobody tell you, though, that there is no "goodness" in beer. There is. It's why the Germans call it liquid bread.

The Beer Belly Myth

Which straight away will lead plenty of cynical folk to a "gotcha" moment (in their minds): beer makes you fat. Everyone has heard of a beer belly. Nobody gets a wine belly (or so they believe). In a word: rubbish. It is a matter of "calories in" and "calories out." If you burn off more calories than you put into your body, you will not accumulate fat. However you *must* count the calories, every last one of them, including those from your preferred beverage.

And the main source of calories in any alcoholic beverage is alcohol. For that reason, the more potent the brew, the bigger impact on the mind, but also on the midriff, unless you are counting those calories and ensuring that you exercise them away. This is the very reason why an English cask ale of say 3.7% ABV features fewer calories than, say, a German bock.

There are some in my own family who eschew beer because they say it is more fattening than is wine. So ponder this. There's me and my relative companionably chatting about soccer in a Cheshire pub. I am delighting in my two standard drinks in the shape of a solitary pint of delicious draft bitter, and he is taking his in the form of sipping a large glass of pinot noir. I am supping about 170 calories; he is at exactly the same number. Factor in that, side by side, I am faring the better in terms of gaining some beneficial nutrients from the beer that are not present in the wine. Then I know who is making the wiser selection.

None of this is to say that I am necessarily healthier than my family drinking companion. Indeed I fully appreciate that by any visual criterion he is the healthier. And I know why. He does half marathons, even full ones on occasion. I don't even run 10 yards. He walks the Pennine Way and the Lyke Wake Walk. I walk to the restaurant.

It is all about lifestyle. It was captured in a curious research study by Grønbæk and colleagues in Copenhagen. They parked themselves outside a supermarket and asked emerging customers for their shopping receipts. Three and a half million of them. They divided them up into those who had been buying wine and those who had been buying beer. The abstract of the paper states that "wine buyers bought more olives, fruit and vegetables, poultry, cooking oil, and low fat cheese, milk, and meat than beer buyers. Beer buyers bought more ready cooked dishes, sugar, cold cuts, chips, pork, butter or margarine, sausages, lamb, and soft drinks than wine buyers." In other words, wine drinkers seem to make healthier food selections.

What anyone should strive for is a well-balanced diet as part of a well-balanced lifestyle. Everything in moderation. And that includes other foodstuffs that tend not to attract the same vitriol as does beer.

What Else Do You Enjoy and Have People Scaring You Over?

The World Health Organization observes that eating processed meat is at your own risk. The equivalent of two rashers of bacon a day or wolfing a hotdog increases the chances of you getting bowel cancer by 18%. Will that keep me away from a full English breakfast in an olde worlde English hotel or relishing a hotdog in transit through Chicago O'Hare? Certainly not. I will take my chances, just as I will with my beer, moderately consumed.

Chocolate can cause relaxation of your lower esophageal sphincter. Impact? Your stomach contents head north and you get that unpleasant burning sensation from the acid. Does it make me want to avoid a bar of Cadbury's Fruit and Nut? Certainly not—but I won't wolf down more than one bar. Just as I will display moderation with my beer.

It has been claimed that heavy coffee consumers are at least four times more likely to get a heart attack and have a greater risk of getting diabetes. Will that stop me using my Starbucks Gold Card? I fear not. Just as I will try not to be an excessively heavy consumer of either this beverage or beer.

Cheese can contain a lot of sodium, which may increase your blood pressure and the attendant risk for stroke and heart disease, as well as problems with the kidneys. It suits me—as I can think of nothing worse than Welsh rarebit, for instance. I do like a bit of Lancashire or red Leicester though, to nibble on alongside my beer. Both in reasonable quantities, naturally.

We can carry on with pretty much any food, but I think you are getting the drift. All things taken temperately (if I am permitted to use the word).

And so we have seen that it is a veritable minefield if one tries to draw correlations between the consumption of any one type of dietary component (e.g., beer) and its impact on health. This applies whether you are considering the negative or indeed the positive.

It rather seems, though that the inherent tendency of the medical professions and the news services that cover them is to stress the downside of alcohol consumption. It would be remiss of me to fail to draw attention to the diverse studies that have correlated moderate alcohol consumption to an improved state of health. And I feel sure that the responsible medics and scientists who delivered their work through peer-reviewed publications would resent its being confined in the category of "old wives tales."

Beer Is Good For You

There are numerous papers relating moderate alcohol consumption to the reduced risk of coronary heart disease, the argument being that there is a lowering of the accumulation of bad cholesterol blocking the arteries. Considering that this disease is the biggest killer in the United States, it is at the very least worth pondering.

We might also turn to an interesting study that did not rely upon drawing correlations based on human behavior. This was performed by Jan Hlavacek and colleagues from the company that brews Pilsner Urquell, using some very fortunate rats and hamsters. First of all they found that, given the choice between water, a solution of alcohol in water, and beer, the rats preferred the beer. Go figure. In fact the same beers that humans preferred also found favor with the rats. Hmm. But then the authors started to treat their hamsters to some beer or red wine and found that the beer was just as good as the wine in cutting down the levels of cholesterol accumulating in the blood vessels of our furry friends.

In other studies, moderate beer consumption has been linked, among other things, to reduced incidence of late-onset diabetes, of kidney stones, and of Alzheimer's disease.

In some ways most intriguing of all, there have been several studies reporting a relationship of restrained beer consumption to cognitive performance, especially in the elderly. The late professor Robert Kastenbaum advocated a moderate intake of alcohol daily by the elderly to enhance their mental well-being and contentment.

Dr. Mathias Benedek, from the Department of Psychology in the University of Graz, concluded that a pint of beer for men (a little less for women), raised performance scores by up to 40% when they were given a range of creative tasks.

A final point in this general discussion of the relationship between beer and health. And that is that every human is distinct. Our genetic makeup and therefore the ins and outs of our metabolism vary between us. It is now very much the norm to talk about personal nutrition: configuring the content of our diet in relation to what is optimum for our own body. I rather doubt that we will get to the situation where we will be able to say things like, "You, sir, are configured to be best suited to a *Hefeweizen* and you, ma'am, will find that a *Doppelbock* will work very nicely with your metabolism." I am merely observing that different people will react very differently to beer and its components, with some being more affected in either beneficial or, alternatively, in disadvantageous ways by consuming a certain beer (or indeed any foodstuff).

An example close to home. My wife Diane knows absolutely to avoid any beverage that has been stored in wood. A mere sip will give her the most horrendous migraine. There are other components in beer (just as there are in any other dietary item) that some people are more sensitive to than others.

One example is gluten. Although it has become something of a (hopefully transient) food fad to eschew the presence of gluten and gluten-like proteins in the diet, it is certainly the case that there are plenty of people who are sensitive to such proteins, including those with celiac disease. Thus, for the longest time folks with these unfortunate conditions are advised to avoid beers because most of them are made from barley and wheat, both of which contain these problematic entities. It is a sensible precaution, though the actuality is that huge proportions of the undesirable proteins are removed during the malting process and throughout brewing, especially at the mashing stage. There are plenty of famous brands on the supermarket and liquor store shelves that have very little of the unwanted proteins surviving.

However if a brewer deliberately seeks to make a beer devoid of these materials, there are two potential approaches. The first is to use a grist that does not contain gluten-type proteins and make the beer from the likes of sorghum, buckwheat, and millet. Alternatively, a regular type of grist can be used, but an enzyme that chews up the unwanted protein is added to the fermenter. The latter approach leads to products that in the United States cannot be called gluten-free, but rather use wording such as "processed to remove gluten" on the label.

Bias?

I will conclude this chapter with a statement for all those readers who would basically say that you shouldn't trust a darned word I say because I am part of the brewing industry. I would simply make the analogy of another key part of my life, namely my involvement in soccer. I have been writing about soccer since 1985. I have my favorites, namely Wolverhampton Wanderers FC, and I have followed them through some thick and an awful lot of thin times in some 60 years. I can recognize (and speak truth to) their strengths and weaknesses, without fear or favor, and that is because they don't pay me. No brewing company has ever paid me to do research on, or write about, beer and health. Indeed there are some pretty significant companies that prefer that I don't, I think because they are concerned that it will whip up the naysayers and the neo-prohibitionists to ever more extreme negativity. Me? As a scientist I want the truth to emerge through the pursuit of good science and fair-minded and accurate interpretation. We need to deal in facts and not politically driven high-financed policies unfounded in firm realities. And it sure helps to know something about beer if you are going to make pronouncements about it. Heck, you would not want a baseball commentator to grind the dust finely on basketball, would you?

9

Business

We tended to strut a bit in Bass. We were so proud to be part of the UK's biggest brewing company, honored to wear the red triangle on our clothing. The red triangle was the first trademark to be registered as part of the UK's Trade Marks Registration Act of 1875.

What few in the company took much effort to own up to was the history of the company. Bass PLC got to where it did by a remarkable series of mergers, acquisitions, and closures.

Let's start in 1744, with George II on the throne and war declared between Britain and France and fought over in North America. William Worthington established his brewery on High Street in Burton-on-Trent. There were already three other brewers in the town for a population of a very few thousand people. In 1777 William Bass turned from haulage to brewing. Both companies focused primarily on the export trade, via the Trent Navigation and the port of Kingston-upon-Hull but also via train to the burgeoning port of Liverpool. The legend is that a shipwreck in 1827 led to a consignment of the Burton beer being auctioned in the Lancashire city and a reputation grew for the beer in the domestic market.

In the 1840s the Midland Railway, headquarters Derby, came into being. Michael Bass became a director and ensured that the cellars at St. Pancras Station in London were suitable for holding large quantities of Bass ale, in the form of hogsheads (1 hogshead = 1.5 UK barrels = 245.5 liters). By 1877, the town of Burton was brewing almost a million barrels of beer each year. The bulk was produced by Bass Worthington, which arose from a merger in 1826. Then in 1861 there was a further merger with Mitchells and Butlers of Birmingham, a company that already had acquired the likes of Highgate in Walsall and Butlers of Wolverhampton.

Meanwhile Charrington had become a brewer of note at Mile End in London. The company fancied building a brewery in Burton in 1871, to take advantage of the famous hard water there, but retained its London headquarters as an independent company until 1962, when it joined the Northern

United Breweries conglomeration of 16 breweries, the latter being the handi-work of Canadian Eddie Taylor.

Edward Plunket Taylor (born 1901 in Ottawa) was spawned into wealth and inherited his grandfather's Brading Brewery, which he merged with 20 other breweries to form the world's biggest brewing company, Canadian Breweries Limited. With international aspirations, he entered into a deal with the Hope and Anchor company in Sheffield to make his Carling Black Label (these days the biggest selling beer in the United Kingdom), while the Jubilee sweet stout brand made the reverse trip.

Taylor was not content with the sales performance for his lager—so he ac-quired the Sheffield Company and a collection of others, leading to the afore-mentioned Northern United Breweries.

Bass, Mitchells, and Butlers themselves were in merger and acquisition mode, buying a series of companies and closing the breweries of those enti-ties in due course, while of course keeping their pubs and the viable brands.

In 1967, Taylor (whose real love was breeding thoroughbreds) made a bid for Bass. The result was a merger of Bass, Mitchells, and Butler with Charrington, to form Bass Charrington, a company brewing 20% of the UK beer and selling much of it through 11,000 pubs. Brewery rationalization continued.

All told, the Bass Charrington company grew out of 81 takeovers and the merging of 273 brewing entities.

To us (at the time I was with the company between 1983 and 1991) we were just Bass PLC. The names Worthington, Charrington, Mitchells and Butlers (M & B) still featured, as did Tennent Caledonian, our Scottish wing. But there was much besides, notably betting shops (Joe Coral) and hotels (Crest).

And then the Iron Lady marched in.

Margaret Thatcher, the prime minister, spent rather too much time talking to Ronald Reagan. One can imagine how the conversation went.

"Say, Maggie, you are letting those big brewers of yours off far too easily."

"How so, Ronnie?"

"Letting them control everything from brewery to sale. After Prohibition we stuck it to our guys by introducing the three-tier system: brewer, distrib-utor, and retailer. In short, if you brew lots of beer, you can't own a pub or a liquor store selling that beer."

Quite whether Mrs. Thatcher asked the president why this was such a good idea is something I will forever wonder about. The reality is that the big brewers in the United States still have a large slice of control, as they pretty much dictate what gets onto a distributor's truck and thus onto the shelves. In any event she came back to Blighty and drove forward the Beer Orders in 1989.

The basic premise was that the six biggest brewing companies in the United Kingdom (Bass, Allied, Whitbread, Grand Metropolitan, Courage, and Scottish & Newcastle) had a monopoly in that they owned a very large number of pubs and thus had a firm control of the beer supply.

Even allowing for the consolidations of the type described earlier that led to the growth of Bass and which had also occurred with the other large entities, there were still plenty of smaller brewing companies in the United Kingdom and plenty of choice.

Nope, Maggie said. If you own more than 2,000 pubs and you also own at least one brewery, you have to sell all the surplus pubs. However, if you own no pubs you can brew limitless quantities of beer. If you don't brew any beer, you can own as many pubs as you want. The result: all the Big Six got out of brewing and went into pubs and hotels. You make much bigger profits at the retail end than you do at the wholesale end. Bass became Intercontinental Hotels (think Crowne Plaza, Holiday Inns, Indigo, and more), which is, as I write, the second biggest hotel company in the world after Marriott.

The Bass breweries were divided between Interbrew in Belgium and Coors from Golden, Colorado. In due course Interbrew merged with Ambev from Brazil to become Inbev and after a while acquired Anheuser-Busch to become Anheuser-Busch Inbev. Meanwhile Coors merged with Molson to form Molson Coors.

The second biggest brewing company globally after AB Inbev was SAB-Miller, which had arisen from the purchase of Miller from Milwaukee (owned by the tobacco giant Philip Morris) by South African Breweries. In 2016 AB Inbev acquired SAB-Miller. The company was, of course, obliged to offload brands as it truly would have had even more of a monopoly than it enjoyed already. What it really wanted was to get control of South Africa. So it kept the SAB part and sold the Miller brands to Molson Coors. Other brands like Peroni and Grolsch went to Japanese company Asahi.

When it comes to the brands that I knew and loved from Bass, then our biggest seller (and still the biggest brand in the United Kingdom) Carling Black Label, went to Coors. But Bass (with its famed red triangle) went to Anheuser-Busch, thus now AB Inbev, whose other brands include Labatt, Becks, Pilsner Urquell, as well of course as Bud, Bud Light, and Michelob and very many more besides.

We are not at an endpoint on all this, and this book is likely to be out of date in this respect even before it hits the bookshelves. And for sure there are those who rail against this monopolization of beer. Pause to think, though, about some of the other preferences you have in your life and whether you are enjoying the products of mighty organizations that in some cases rather dictate the industry with products that are integral to many people's lives. Think Microsoft. Unilever. Google. Facebook. Monsanto. Chevron. Starbucks. The list goes on.

Successful companies make great products. They have to or they would not survive.

We remain grateful, however, for choice. And there is now tremendous opportunity in the world of brewing. Since President Jimmy Carter signed into law the legislation rendering home brewing legal in the United States in 1978, there have been two great growth surges in the establishment of new brewing companies. As I write there are some 7,500 breweries in the United States, compared with just 89 when Carter put pen to paper. Some don't survive. But the ones with a good business model and with good beer do.

Some are tiny, some are much larger. The Brewers Association definitions are these:

Brewpub: a restaurant-brewery that sells 25 percent or more of its beer on site. The beer is brewed primarily for sale in the restaurant and bar. The beer is often dispensed directly from the brewery's storage tanks. Where allowed by law, brewpubs often sell beer "to go" or distribute to offsite accounts. Note: BA re-categorizes a company as a microbrewery if its off-site (distributed) beer sales exceed 75 percent.

Microbrewery: a brewery that produces less than 15,000 barrels (17,600 hectoliters) of beer per year with 75 percent or more of its beer sold off-site. Microbreweries sell to the public by one or more of the following methods: the traditional three-tier system (brewer to wholesaler to retailer to consumer); the two-tier system (brewer acting as wholesaler to retailer to consumer); and, directly to the consumer through carry-outs or on-site taproom or restaurant sales.

Regional craft brewery: An independent regional brewery with a majority of volume in "traditional" or "innovative" beer(s).

Regional brewery: A brewery with an annual beer production of between 15,000 and 6,000,000 barrels.

Large brewery: A brewery with an annual beer production over 6,000,000 barrels.

As I write, the 10 biggest craft brewing companies in the United States are Yuengling, Boston, Sierra Nevada, New Belgium, Gambrinus, Duvel Moortgat, Bells, Deschutes, Stone, and Oskar Blues.

Some prominent names drop off the list over time, as they become acquired by the bigger global companies. Examples are Lagunitas (Heineken), Ballast Point and Funky Buddha (Constellation Brands—which owns Robert Mondavi wineries and Corona/Modelo), Anchor (Sapporo), Goose Island, Elysian, 10 Barrel, Breckenridge, Blue Point, Four Peaks, Golden Road, Devil's Backbone, Karbach, Wicked Weed, and Shock Top (AB Inbev), and Blue Moon, Leinenkugel, Saint Archer, Hope Valley, Revolver, and Terrapin (Molson Coors). For some consumers this is a deal breaker. Souls sold to devils. The reality is the beer is probably better than it has ever been. It all becomes a personal matter (rather like veganism, religious faith, blind loyalty to a political party, no matter the outrageous behavior of some at the helm of the "club").

This craft phenomenon is not restricted to the States. It is happening across the world. In the United Kingdom, for example, there are some 2,000 brewing concerns. The landmass of the United Kingdom is 93,628 square miles. The landmass of California is 163,696 square miles. Currently there are some 900 breweries in California. At the UK rate, you would expect that number to be almost 3,500 breweries. When people here in California ask me if we have reached saturation point for breweries, I say, "Hell no!"

I say again, though, there has to be a great business model and great beer. You have to brew and sell enough beer to meet payroll and invest. And that beer must be desirable, so that people will make repeat purchases. And you have to be able to get it into the market. It's too tough for most smaller brewers to get onto a distributor's truck—so much competition and so much power still held by the enormous brewing conglomerates. It is a question, then, of having desirable and comfortable facilities to entice people to visit. Be family friendly. Serve food—perhaps a food truck.

Those entering into the business need to recognize that it is not sexy. The bulk of the workload in brewing on the smaller scales includes humping loads and endless cleaning. And neither is the satisfaction likely to be in the paycheck. The majority of brewery employees on the smaller scales are certainly not in it for the money. The romance is in the end product.

10

Onward

I have a lot of time for Glyn Phillips and David Mathis. They both developed excellent brewing companies, respectively Rubicon in Sacramento and American River in Rancho Cordova. They both went under, to the considerable distress of many people, including myself. It wasn't the beer. It was the competition.

The simple truth is that there are a great many folks competing for customers' drinking dollars (or pounds or yen or euros or any other currency you would like to nominate). And I am not only talking brewers fighting for drinkers, with the burgeoning number of brewing companies sprouting up everywhere. I really can't remember how many brewing companies there were in the Sacramento region when I first came to live in the region in 1999. Maybe there were half a dozen. As I write there are around 75. The same surge is happening across the nation. Great for customers if you are seeking to hop around tasting opportunities. But the competition is immense.

Brewers are also up against wine and cider (now being produced by many a brewer), gin (which has been reborn), and diverse other alcoholic beverages.

It gets worse. Younger drinkers are increasingly eschewing alcohol and turning more to other fluids. As we all know coffee is sexy—the right way to cold brew, frequently incorporating nitrogen to smooth out those harsher notes just as per Guinness. Diverse teas, including those with beads that are sucked through wide bore straws.

It's dog eat dog out there. Which perhaps explains the encroachment of the wacky fringes and bizarre ingredients, higher and higher alcohol contents, and of course derivative products like the malternatives and alcoholic seltzers.

Any brewing company that seeks to thrive in the current marketplace must have a strong new-product development program. It must keep an eye on what the competitors are doing. A classic example would be hazy beers. Hoary old traditionalists like me are not fans. But there are plenty of beer drinkers who are captivated by them, so there are few brewers who avoid the juicy IPA sector, for they would be excluding themselves from one of the most successful product types of recent years.

It is often said that one does not drink what one's parents did. I guess if we can get personal and generalize on a generational basis, I know full well that many of the vast number of students and young brewers that I have taught or guided over the years are far more prepared to embrace the extreme fringes than I am. For example, I know that some of them would have no difficulty whatsoever (law permitting) combining marijuana and beer in a single beverage, whereas to me that would be utterly unacceptable.

I have written plenty in this book already about the merits of beer. I have noted also the downturn in beer consumption across the world, including in the traditional beer-consuming countries such as Belgium, the Czech Republic, Denmark, and the United Kingdom. It is, of course, rather more complicated, in that there remains a lot of interest in beer, but perhaps of lesser volumes of different types of beer. Furthermore there is interest in beers from other nations—thus the domestic market for beer in a country like Belgium is in decline but there are increasing amounts of beer brewed for export from that country. In the United States there is of course an ongoing growth in the so-called craft sector and in imports, but the famous long-standing lager brands are hurting. Small wonder that the big guys are buying up the smaller entities.

Let us simply accept that there will always be changes in tastes and enthusiasms. It is hard to fight them, so let us take the stance that we should let new fads have their moment in the sun. Gin is a great example. In the middle of the 18th century it was called Mother's Ruin on account of it being a low-tax, dubiously produced beverage that corrupted and defiled the women (and men) of cities like London. Nowadays it is totally sexy. For heaven's sake, people have even glamorized the tonic to mix with it. Gin distilleries are springing up everywhere. It's okay. I like my gin and tonic. I like too, my whiskey and water. I am known even to have the occasional Southern Comfort. And of course I am a rapacious consumer of tea and coffee.

My point is that I compartmentalize my beer separate from these other libations. I have my preferences and expectations for my beers in just the same way as I have a range of favored foods. I know a good Indian curry when I taste it. Ditto Thai food, and pizza, and shellfish, and any other food you care to nominate.

All of these dishes and drinks form a part of my dining and drinking delectation. Sometimes it's one dish. Sometimes it's another. The same with drinks. But (and maybe I lack adventure) I don't mix and match. I never have bacon with clams for breakfast. I don't savor treacle pudding with gravy.

Trusting that you have remained with me in this flight of fancy, I hope you are realizing my point that we might fairly have expectations when it comes to beer. It need not be boring. The scope is immense and, frankly, there is no need for the outer limits of reason in terms of what it really is sensible to use to make beer and what is responsible in terms of alcohol content.

I believe that there is much to be done in the field of education to better serve the cause of beer. It has become far too easy for brewers to try to attract attention and dollars by going to extremes. I am convinced of the merits of giving customers a far better appreciation of what beer is and why it is the sensible drink to be purchasing (enjoyment, food pairing, health, etc.). I would also go as far as to say that there needs to be legislation about what is and what is not a legitimate component for a beer to be labeled as such.

Note that I am not saying in this that a company should not be able to make any number of products in a brewery. I am very much for any brewing company maximizing its performance, and I have no right or authority to declare what it should or should not do (although I retain the right to know what I personally like and dislike). I am merely saying that it is only those within certain guidelines that should be able to be labeled as beer.

There are already examples of this. Let's go to Japan. They don't grow malting barley in Japan, so it needs to be imported. Presumably to protect domestic crops, a tax advantage is in place which means that if a product is made using less than 25% malt, it attracts a far lower duty than regular beers. These drinks cannot be called beer (they are called *happoshu*), and they are made using significant amounts of adjuncts, but they look like beers, are packaged like beers, and are on the shelves in the shops alongside beers. The difference is that they are retailed at a much lower price. Guess what the Japanese have gravitated to buying? There is a third category, drinks made with zero malt and made from stuff like pea starch and soybeans and which attract even less duty than *happoshu*. They are cheapest of all. To *my* palate they are substantially inferior to the time-honored Japanese beers—but the relatively low cost make them attractive.

Here is a direct impact of legislation. Is it too hard to imagine a reverse situation, one where traditional beers made within certain guidelines (perhaps not as extreme as the Reinheitsgebot but which nonetheless allow for the use of long-standing adjuncts such as rice, corn, candi sugar, and so on) attract a lower rate of taxation than those that deviate from the norm in some way?

I hope I will not lose the friendship of many of my long-standing buddies in suggesting that there is something to be said for taxation in proportion to alcohol concentration. Remember that in the United Kingdom, duty is levied at incrementally (0.1% at a time) increasing levels as the ABV. I feel that there is tremendous scope for beers at the lower end of the ABV range in countries like the United States. I say again that I am no fan of alcohol-free beer, but there is much to be said for beers in the 2%–4% ABV range.

Just recently I was giving a talk to a home brew club in El Dorado Hills, California. I was invited to choose a beer from the board in the pizza place. I think the lowest ABV was around 6% ABV. My home is 40-plus miles away from that city. No way did I want to have more than a solitary glass of the weakest (sic) ale on offer. Had I been able to have enjoyed a couple of leisurely pints at 3% ABV I would have enjoyed my evening rather more.

These are exciting times in many ways. There are people who are interested in trying beer nowadays who most definitely were not beer drinkers before. There are plenty of people who savor their journeys to different tasting rooms, sampling a rich diversity of beers. My thesis is that there could be an even greater selection covering a far wider range of strengths (within sensible bounds) and with a huge span of flavors by sticking to long-standing ingredients. We should focus on demonstrating how this vast array of ales and lagers can pair sublimely with any food you care to nominate. We need to speak truth to the reality of the health pluses and minuses associated with beer drinking. We should strive to carve out from the beer world the stupidity and the irresponsibility, and we should celebrate beer for what it is: the original basis for static civilization and the savior of humankind.

Further Reading

If you are looking for a reference book that covers most aspects of beer with a fair degree of authority, then I suggest *The Oxford Companion to Beer*, edited by Garrett Oliver.

For those of a more scientific persuasion who seek to grind the dust fine on the underpinning science and technology of beer production, my ego does not prevent me from recommending my own *Scientific Principles of Malting and Brewing* (American Society of Brewing Chemists).

The same publishing house produced my five books (so far) in a series dedicated to key aspects of beer quality *Foam, Flavor, Freshness, Color & Clarity* and *Quality Systems*. They are akin to the manual you get with your car: what to do if things go wrong and how to ensure that things don't go wrong. You can actually take online classes built around these books (https://extension.ucdavis.edu/subject-areas/beer-quality-series).

Turning to history, then, I recommend *A History of Beer and Brewing* by Ian S. Hornsey (Royal Society of Chemistry) and *Ambitious Brew: The Story of American Beer* by Maureen Ogle (Harvest Books).

There have been a number of books on beer styles. Perhaps you might like *Beer Styles from around the World* by Horst Dornbusch (Cerevisia Communications). I also strongly recommend *Belgian Beer, Tested and Tasted: The Complete Guide* by Miguel Roncoroni and Kevin Verstrepen (Lannoo). Yes, it focuses on beers from that wonderful country. But the way in which these authors present information on these beers and the principles they apply are applicable to beers from anywhere, and they will greatly help serious beer students interpret their sensory experiences.

When it comes to beer and food pairing, maybe you should seek out Julia Herz's *Beer Pairing: The Essential Guide from the Pairing Pros* (Voyageur Press). In terms specifically of cheese, you need to devour *Cheese & Beer* by Janet Fletcher (Andrews McMeel).

The business of beer is addressed by Johan Swinnen and Devin Briski in *Beeronomics: How Beer Explains the World* (Oxford University Press), while if you want to be truly inspired and realize just how hard it is to develop a truly great brewing company then you really must buy *Beyond the Pale: The Story of Sierra Nevada Brewing Co.* by Ken Grossman (Wiley).

Finally you might feel tempted to delve into a few of my other offerings: *Beer: Health and Nutrition* (Wiley-Blackwell), *Beer Is Proof God Loves Us: The Craft, Culture, and Ethos of Brewing* (FT Press), and *Grape versus Grain* (Cambridge University Press).

Finally, if you prefer to listen, you might try the *Brewmaster's Art* (www.recordedbooks.com/title-details/9781440715419).

Glossary

abbey beer. Beer in the style of those brewed by the Trappists, but from breweries that are not associated with that religious order.

accelerated fermentation. Fermentation carried out under conditions where it proceeds more rapidly, e.g., by operating at a higher temperature.

acetaldehyde. A substance giving the aroma of green apples to beer and usually associated with incomplete yeast activity or with contamination by the bacterium *Zymomonas*.

acetic acid. The acid associated with vinegar that, if present in detectable quantities in beer, is a sign of contamination with acetic acid bacteria.

acetic acid bacteria. Organisms that produce acetic acid. They need oxygen and so are only a problem in beers that encounter significant amounts of oxygen, such as cask-conditioned ales.

acidity. The extent of acid presence in beer. All beers are acidic (pH less than 7) and most are in the pH range 4 to 4.5. The lower the pH, the higher the acidity.

acidulated malt. Malt produced with a deliberate acidification stage in the malthouse through the action of lactic acid bacteria populating the grain. Such malt will lead to a lower pH of wort produced in the brewhouse.

additive. Agent used to aid the malting and brewing processes or the stability of the finished beer. Many brewers eschew the use of these.

adjunct. A source of fermentable extract other than malt for use in brewing.

aftertaste. The lingering taste and aroma of beer detectable after the initial tasting of the product.

aging. The holding of beer in order for it to be converted to the desired state (flavor, appearance) for retail to the consumer.

alcohol. A class of organic compounds (substances containing the carbon atom other than carbon dioxide [and related compounds such as carbonates] and carbon monoxide). The principal product of fermentation by yeast is ethyl alcohol (ethanol). Other ("higher") alcohols are also produced in much lower quantities by yeast, and they are implicated in the flavor of beer and as the building blocks of the more flavor intense esters.

alcohol by volume (ABV). The amount of alcohol in a beverage in which the ethanol content is quantified in terms of volume of ethanol per volume of beverage (milliliters of alcohol per 100 mL of beer).

ale. A type of beer produced using a top-fermenting (ale) yeast. In medieval England, "ale" referred to unhopped beer.

alpha acid. Resin from the hop that is the precursor of the bitter compounds in beer.

American Malting Barley Association (AMBA). A non-profit organization that coordinates activities dedicated to maintaining and developing malting grade barley in North America.

American Society of Brewing Chemists (ASBC). A professional body especially for scientists and analysts working in the brewing and related industries. It has a range of publications, including a compendium of standard methods.

antifoam. A material added to fermentations to suppress excessive production of foam, thereby maximizing the extent to which a vessel can be filled.

antioxidant. A material either native to a raw material or else added that protects against the damaging influence of oxygen.

aphid. Insect that spoils batches of raw materials.

apparent extract. An imprecise measure of the extent to which fermentation has occurred. It is derived by taking the specific gravity of the finished beer. This specific gravity is increased by the presence of residual sugars and other substances surviving from the wort but is lowered by ethanol. Although a rough idea of the extent of fermentation can be gauged by comparing this value with the specific gravity of the wort prior to fermentation, a more reliable index is to use the real extract value (see **real extract**).

aroma. The smell of beer and everything else associated with the production of beer.

aroma hop. Hop variety said to give particularly prized aroma characteristics to a beer. Also called noble hop.

Aspergillus niger. A fungus that produces the enzyme glucoamylase, which can convert non-fermentable dextrins into fermentable sugars. Features prominently in the production of sake.

astringency. A drying of the palate.

ATP bioluminescence. A technique for detecting microorganisms and soil (dirt) by measuring the amount of light produced by the action of the enzyme luciferase acting on ATP present in the sample.

attenuation. The extent to which fermentation has occurred, gauged by comparing the starting specific gravity with the final specific gravity.

auxiliary fining. Agent used alongside isinglass to facilitate the settling of insoluble materials from green beer.

bacteria. Single-celled organisms that for the most part spoil beer and its raw materials but which can sometimes be desirable, e.g., in the production of sour beers.

balling. A measure of the percentage sugar by weight in wort. Essentially the same as degrees Plato (or Brix as used in the making of wine).

barley. *Hordeum vulgare* or *Hordeum distichon*; a member of the grass family and the principal raw material for malting and brewing worldwide.

barley wine. A very strong type of ale of long standing in England.

barrel. A volume measure of beer (= 31 US gallons, 1.1734 hectoliters; = 36 UK gallons, 1.6365 hectoliters).

barrel aging. Storage of beer in barrels, generally oak, to extract flavors directly from the wood and sometimes allow organisms inhabiting the barrel to effect change in the beer. The barrels may have previously been used to store wines or whiskies of various types.

base malt. The relatively lightly dried malt that forms the major grist component for every beer because it provides the enzymes needed to convert non-fermentable starch into fermentable sugars.

beading. The formation of bubbles of carbon dioxide in a glass of beer and their rise to the top of the drink.

beechwood. Shavings that are used in the storage of Budweiser and Bud Light. They represent a surface that can be populated by the yeast important to remove undesirable flavor substances such as diacetyl and acetaldehyde. Conversely, the burning of beechwood delivers the smokiness needed in the production of the malts used for *Rauchbier*.

beer engine. The pumping system employed to deliver cask-conditioned beer from the cellar to the glass in the bar, based on suction.

Beer Judge Certification Program (BJCP). A US-based organization that trains and certifies beer judges.

beer stone. A precipitation of calcium oxalate in beer dispense pipes.

biological stability. The extent to which a beer is able to resist contamination by microorganisms.

bitterness. A flavor characteristic customarily associated with beer; also the term used to quantify the content of bitter compounds (iso-alpha acids) in beer.

bitterness unit. A numerical approach to assessing bitterness, with quantification based on absorbance of ultraviolet light by extracts of the beer. The amount of light absorbed at 275 nm is multiplied by 50 to give the BU (also known as IBU).

boiling. The process of vigorously heating sweet wort at boiling temperatures.

bottle conditioning. Ageing of beer in the final container (applied also in some canning operations) in which residual yeast converts sugar added before packing into carbon dioxide, thereby bringing the amount of carbonation to the level desired in that package.

bottom fermentation. Traditional fermentation mode for lagers where yeast collects at base of fermenter.

breakdown. Deterioration of a beer.

Brettanomyces. A type of yeast named for the British brewers who insisted that it was necessary to bring beer into condition. Although important in the production of certain "wild" beers, it is generally considered to be a spoilage organism these days.

Brewers Association. A Boulder, CO–based organization primarily focused on supporting the craft brewing industry, inter alia running events such as the Craft Brewers Conference and the Great American Beer Festival.

brewhouse. The part of the brewery in which grist materials are converted into wort.

brewster. A female brewer.

bright beer. Beer post-filtration.

bright beer tank. The vessel to which a beer is run after filtration and before packaging. Sometimes called a fine ale tank.

Burtonization. Adjustment of the salt content of brewing liquor to render it similar to that of the water at Burton-on-Trent in England.

butt. A container equivalent to three barrels (UK) of ale.

Campaign for Real Ale (CAMRA). An organization originally established to champion cask-conditioned ale.

candi sugar. An adjunct made from the heating of solutions of beet sugar with salts to provide a range of products with different intensities of color and caramelized flavors. A key addition to the kettle in the production of the high-alcohol Trappist beers.

carbonation. The amount of carbon dioxide in a beer and also the act of increasing the level of carbon dioxide.

carbon dioxide. The gas that provides the sparkle and effervescence to beer.

carboy. Large glass or plastic containers for beer, long since used as small-scale fermenters.

cardboard. An undesirable flavor note that develops in packaged beer on storage.

carrageenan. An extract of seaweed used to aid solids removal in the wort boiling stage. Also known as Irish moss.

cask. The traditional vessel for holding unpasteurized English ale.

cask breather. A valve system linked to carbon dioxide associated with cask ale, allowing carbon dioxide rather than air to replace the space in a cask as it is being emptied using a beer engine.

cask conditioning. Maturation of beer in casks by added yeast converting added sugars into carbon dioxide.

Cask Marque. An organization employing trained experts to visit pubs and other outlets to inform about and audit the handling of cask-conditioned ale.

cell. The basic unit of any living organism.

cellar. The part of a brewery containing the fermenters and the conditioning vessels. Also the part of a retail outlet (e.g., bar) in which the beer containers (e.g., casks) are stored.

charcoal. A material capable of adsorbing flavors and colors from liquids which it contacts. Used, for example, to treat liquor coming into a brewery.

chilling. The cooling of liquid streams in a brewery, e.g., hot wort going to the fermenter, or green beer passing to conditioning and filtration.

chill haze. Turbidity that appears in beer if it is chilled to 0°C (32°F) but which redissolves at 20°C (68°F).

chromosome. The form in which the genetic material of a cell (DNA) is held in eukaryotic cells.

cicerone. A trained individual expert in articulating the presentation and properties of beer.

CIP (cleaning in place). An integrated and automated system of cleaning with caustic and/or acid installed in modern breweries.

clarification. The removal of insoluble materials from beer, including processes such as settling and filtration.

cling. The adhesion of foam to the walls of a beer glass (also known as lacing).

coalescence. The tendency of bubbles in beer foam to merge together and form bigger bubbles.

cold break. Insoluble material that precipitates out of wort on chilling.

colloidal instability. The tendency of a beer to throw a haze on storage.

color. The shade and hue of a beer.

conditioning. The maturation of beer in respect of its flavor and clarity.

continuous fermentation. A process in which wort is converted to green beer in a few hours by passage through a vessel holding yeast.

contract brewing. Use of a brewery to produce a beer brand that is not owned by that company.

conversion. The stage in mashing when the temperature is raised to enable gelatinization of starch and subsequent breakdown of the starch by amylases.

cooker. A vessel in the brewhouse in which adjuncts with very high gelatinization temperatures are cooked.

cooler. A device (often called a paraflow) in which hot wort flows counter to a cooling liquid in order to bring it down to the temperature at which fermentation will be carried out.

coolship. A shallow vessel located close to the roof of traditional breweries, especially those making sour beers, to which the hot wort post-boiling is pumped to slowly cool.

copper. The vessel (often called the kettle) in which wort is boiled with hops.

corn. Maize. The word is also used to describe individual grains of barley.

cropping. The collection of the yeast that proliferates during fermentation.

crown cork. The crimped caps used on beer bottles.

culms. The rootlets of germinated barley that are collected after kilning and sold as animal feed.

curing. The higher temperature phases of kilning when flavor and color are introduced into malt.

cylindro-conical vessel (CCV). A tall fermentation vessel with a mostly cylindrical body, but a cone-like base into which the yeast collects after fermentation.

decoction mashing. Practice originating in mainland Europe in which a mash is progressively increased in temperature by taking a proportion of it out of the mix and boiling it prior to adding it back into the whole. The purpose is to go through a series of temperature stands to allow enzymes of different tolerance to heat to do their job.

descriptive test. Beer tasting protocol in which trained tasters describe the taste and aroma of beer according to a series of defined terms.

dextrin. Partial breakdown product of starch that consists of several glucose units and which is not fermentable by yeast.

dextrose. Glucose.

diacetyl. A substance with an intense aroma of butterscotch or popcorn that is produced by yeast during fermentation but which is subsequently mopped up again by the yeast.

diatomaceous earth. The skeletal remains of microscopic organisms used in powder filtration of beer (also known as kieselguhr).

difference test. Blind tasting procedure in which tasters (including the untrained) are asked to differentiate samples of beer.

dimethyl sulfide. Compound which imparts a significant canned corn-like flavor to many lager-style beers.

dissolved oxygen. The amount of oxygen dissolved in a wort prior to fermentation or, more commonly, the amount dissolved (and undesired) in beer.

distribution. The transportation of beer from the brewery and warehouse to the market.

downy mildew. A disease of hops.

draft beer. Either beer in cask or keg, or sometimes unpasteurized beer in small pack.

drinkability. The property of beer that determines whether or not a customer judges it worthy of immediate re-purchase.

dry beer. Beer genre in which the beverage contains relatively low residual sugar.

dry hopping. Traditional procedure in which hops are added to the finished beer prior to shipment from the brewery.

duty. Excise tax on beer.

dwarf hop. Hop that grows to a lower height than traditional varieties.

ear. The head of a barley plant that holds the grain.

effluent. The liquid waste material produced in a malthouse or brewery.

embryo. The baby plant in the grain.

enzyme. A biological catalyst, comprising protein.

essential oil. The aromatic component of hops.

ester. A class of substances produced by yeast that afford distinctive, sweet aromas to beer.

ethanol. The principal alcohol in beer, which is the major fermentation product of brewing yeast and which affords the intoxicating property to the beverage. Originally called ethyl alcohol.

evaporation. A measure of the water loss during wort boiling.

excise. Tax on alcoholic beverages levied by government agencies.

extract. Liquid material derived from malt or hops that can be used as an alternative to solid materials to afford flexibility or cost savings.

feed grade barley. Barley that yields a relatively low level of extractable material after conventional malting and mashing.

fermentability. The proportion of sugars in wort that can be converted into ethanol.

fermentation. The process by which sugars are converted into ethanol by yeast.

filtration. The clarification of beer (sometimes people refer to the recovery of wort from spent grains as "filtration," but strictly speaking this is "wort separation").

fingerprinting. The differentiation of yeasts (or barleys) by analyzing the pattern of DNA fragments produced from them.

fining. Material used to clarify wort and especially beer by interacting with solid materials and causing them to sediment.

firkin. A beer container holding 25% of a UK barrel.

flash pasteurization. Heating of flowing beer to a high temperature (e.g., 78°C, 172°F) for less than a minute in order to inactivate microorganisms.

flavor profile. An expert semi-quantitative evaluation of beer flavor made by trained tasters using defined taste and aroma descriptive terms.

flavor stability. The extent to which a beer is able to resist flavor changes (usually undesirable) within it.

flocculation. The tendency of yeast cells to associate.

floor malting. Traditional labor-intensive mode of malting in which germination is conducted on a floor with intermittent raking into piles to allow heat buildup or thinning out of piles if cooling is required.

foam. The head (froth) on beer.

foam stabilizer. Either endogenous materials (e.g., proteins from malt) that stabilize foam or materials added to beer to protect foam (e.g., propylene glycol alginate).

fobbing. Forcing foam to come out of a bottle after filling and before applying the crown cork, this to exclude air from the container.

font. The unit on the bar that labels the draft beer being served from that tap.

franchise brewing. The brewing of one company's beer under license by another company.

fungicide. An agent sprayed onto crops such as barley and hops to prevent the growth of fungi thereon and ensure that those crops are healthy, high yielding, and don't introduce any harmful materials into the brewing process.

fusarium. A contamination of barley that can cause a beer made from that beer to gush.

gallon. A standard unit of beer volume (a US barrel = 0.8327 UK barrel).

gelatinization. A disorganization and loosening of the internal structure of starch granules by heating, rendering the starch more amenable to enzymic hydrolysis.

genetic modification. A process of modifying the genome of an organism by introducing specific pieces of DNA from an exogenous source. Not desired by the majority of brewers.

genome. The information code in a cell, held within DNA, which determines the nature and behavior of that cell.

germination. The process by which steeped barley is allowed to partially digest its endosperm and the embryonic tissues to partially grow.

gravity. The strength of wort in terms of concentration of dissolved substances, as measured traditionally using a hydrometer.

gravity dispense. Serving beer from a barrel, usually located behind the bar, using a tap and the simple force of gravity.

green beer. Freshly fermented beer prior to conditioning.

green malt. Freshly germinated malt prior to kilning.

grist. The raw materials (malt and other cereals) that will be milled in the brewhouse. More loosely applied also to those adjunct materials that don't require milling (e.g., syrups to be added to the kettle).

grits. Particles larger than those of flour that are produced in the milling of grain.

growler. A container that is taken to a brewery to be filled with draft beer for consumption away from the brewery.

gruit. Blends of herbs and spices historically used to flavor beer, especially before the ascendancy of hops.

gushing. The uncontrolled surge of the contents of a beer from the package after opening.

gyle. A batch of wort or beer as it progresses through the process.

Hallertau. A famed hop-growing region in Germany.

hammer mill. A mill that grinds malted barley down to extremely fine particles that are suited to a mash filter for subsequent wort separation, but not a lauter tun.

haze. Turbidity.

hazemeter. An instrument for measuring the clarity of beer: operates on the principle that particles scatter light. The more light scattered, the more particles are present. Some haze meters measure the amount of light scattered at right angles (90°) to the light beam shone at the particles. Other meters ("forward scatter" meters) measure the light deflected at 13°. The former type is sensitive to extremely small particles, the latter to big particles.

heat exchanger. Device for rapidly cooling down liquid streams, e.g., boiled wort. The hot liquid flows countercurrent to a cold liquid on either side of thin walls. Heat passes from the hot to the cold liquid.

hectoliter. 100 liters.

high-gravity brewing. Technique for maximizing vessel utilization whereby the wort being fermented is more concentrated than necessary to make the desired strength of beer. After fermentation, the beer is diluted to the required alcohol content.

hogshead. A container holding 1.5 UK barrels of beer.

hop. Plant that provides bitterness and aroma to beer.

hop back. Vessel rarely found these days that was used to separate boiled wort from residual solids by passage through a bed of waste hops.

hop cone. The flower of the female hop plant, which is the part of the plant used in the brewing process.

hop garden. Where hops are grown. Called yards in the United States.

hop oil. The component of hops providing aroma (essential oils).

hop pocket. A large sack packed with hops.

hop preparation. Extract of hops, usually made with liquid carbon dioxide, and used at various stages in the brewing process to introduce bitterness or aroma to wort or beer more efficiently.

hop resin. The precursors of bitterness in beer (alpha acids).

hop utilization. The proportion of the hop resins added in the kettle that end up as bitterness in the finished beer.

hopped wort. Wort after the boiling stage.

husk. The protective layer around the barley corn.

hydrogen sulfide. A substance with the aroma of rotten eggs that is undesirable in most beers though not the cask-conditioned ales from Burton-on-Trent, England.

hydrolyzed corn syrup. Material produced by the acid or enzymic hydrolysis of corn starch and that can have different degrees of fermentability. Added to the wort kettle, thereby providing an opportunity to extend brewhouse capacity by avoiding the need for mash extraction and separation stages.

hydrometer. Device operating on a principle of buoyancy for measuring the specific gravity of a solution: the higher it floats, the more material is dissolved in the solution.

ice beer. Beer produced with a process including ice generation.

imperial. Originally used in the brand names for strong beers brewed for the imperial courts of Europe, notably Russia. Nowadays used generally for some of the more alcoholic brews.

infestation. Condition whereby a raw material in the malthouse or brewery has animal life within it, e.g., insects in badly stored barley.

infusion mashing. Mashing at a constant temperature, classically around 65°C (149°F), primarily to effect the breakdown of starch to sugars.

insecticide. Material sprayed onto crops either during growth or during storage to eliminate insect infestation.

Institute of Brewing and Distilling (IBD). A global membership organization for professionals in or associated with the alcoholic beverages industry. Primarily known for its examination system.

International Bitterness Units. See **bitterness units.**

invert sugar. Hydrolyzed sucrose (i.e. a mixture of glucose and fructose in equal proportions).

iron. Inorganic element that can enter into beer from some raw materials (e.g., filter aids) and potentiate oxidative damage.

isinglass. Preparation of solubilized collagen from the swim bladders of certain fish, used for clarifying beer; normally referred to as "finings."

iso-alpha acid. Bitter component of beer derived from hops.

isomerization. The conversion of hop alpha acids into iso-alpha acids, achieved during wort boiling.

jetting. Firing a small amount of sterile water into newly filled bottles to induce a foam to rise and expel air prior to capping.

keg. Large container for holding beer, for subsequent draft dispense by pump.

kettle. Brewhouse vessel in which wort is boiled; also known as a "copper."

kieselguhr. Mined powder, derived from skeletons of microscopic animals (diatoms), used to aid the filtering of beer. Also known as diatomaceous earth.

kilderkin. A container containing half a UK barrel.

kilning. Heating of germinated barley to drive off moisture, and introduce desired color and flavor.

kosher beer. A beer compliant with kashruth, the Jewish dietary rules. Beers produced from malt, hops, yeast, and water are generally acceptable without certification, but many people insist on this certification.

lacing. Tendency of beer foam to stick to the side of the glass (also known as cling).

lactic acid bacteria. Bacteria that produce lactic acid and a range of other flavor-active components.

lager. A type of beer produced by bottom-fermenting yeast and traditionally produced in a relatively slow process, which includes lengthy cold storage ("lagering"). The word "lager" is derived from the German "to store."

lagering. See **lager**.

large pack. Kegs or casks.

late hopping. Practice of adding a proportion of the hops very late in the wort-boiling phase in order to retain certain hop aromas in the ensuing beer.

lauter. The act of separating sweet wort from spent grains, and also the vessel used to perform this duty.

light beer. Beers in which a greater proportion of the sugar has been converted into alcohol.

lightstruck. Skunky flavor that develops in beer exposed to light.

liquid carbon dioxide. Solvent produced by liquefying carbon dioxide gas at low temperatures and high pressures; used for extracting materials from hops.

liquor. Water.

Lovibond. A color measurement strategy named for Joseph Lovibond.

low-alcohol beer. A beer containing a low level of alcohol (e.g., less than 2% ABV, although the definition differs between countries).

lupulin. The glands in hop cones that contain the resins and oils.

maize. Corn.

malt. Dried germinated cereal, mostly from barley and wheat.

malting. The controlled germination of barley involving steeping, germination, and kilning so as to soften the grain for milling, develop enzymes for breaking down starch in mashing, and introduce color and desirable flavors.

maltings. The name for a malthouse in the United Kingdom.

malting grade. Score allocated to a barley variety that indicates whether it will give a high hot-water extract after conventional malting and mashing.

Maris Otter. A winter variety of malting barley from England that is highly prized, including as malt produced by old-fashioned floor-malting approaches.

masher. Device positioned before the mash mixer that facilitates intimate mixing of milled malt and hot liquor.

mashing. Process of contacting milled grist and hot water to effect the breakdown of starch (and other materials from the grist).

mash filter. Device incorporating membranes for separating wort from spent grains.

mash tun. Vessel for holding a "porridge" of milled grist and hot water to achieve conversion of starch into fermentable sugars.

mashing off. Conclusion of mashing, when the temperature is raised prior to the wort separation stage.

Master Brewers Association of the Americas (MBAA). Professional body for brewers, notably those resident in North and South America.

maturation. The post-fermentation stages in brewing when beer is prepared ready for filtration.

metabolism. The sum of the many chemical reactions involved in the life of a living organism such as barley or yeast.

microbes. Microorganisms, including bacteria and yeasts.

micronized grains. Cereal grains heated intensely using infrared to swell them. The grains soften in this process. They can be used as adjuncts. The process originated in the production of breakfast cereals, e.g., puffed wheat.

milling. The grinding of malt and solid adjuncts to generate particles that can be readily broken down during mashing.

mixed-gas dispense. Driving beer from containers (notably kegs) using a mixture of carbon dioxide and nitrogen gases, the latter causing a much more stable foam.

modification. The progressive degradation of the cell structure in the starchy endosperm of barley.

moisture content. The amount of water associated with materials such as barley, malt, hops, or yeast.

mold. Infection of barley or hops.

mouthfeel. The "tactile" sensation that a beer creates in the mouth (also referred as texture).

near beer. Non-alcoholic beer permitted to be produced during Prohibition in the United States.

New Albion. The first commercial craft brewery established in the United States following the law allowing home brewing. It was made in Sonoma, California, by Jack McAuliffe.

Ninkasi. Goddess in Sumeria who is considered the mother of all creation, including beer.

nitrogen. There are two completely separate meanings for nitrogen in malting and brewing: (*a*) the nitrogen atom as it is found in proteins (therefore its level in barley, malt, or wort is a measure of how much protein they contain); (*b*) gaseous nitrogen (N_2), which is sometimes introduced into beer to enhance foam. This process of introduction is called nitrogenation.

noble hops. Aroma hops.

non-alcoholic beer. Beer containing very low levels of alcohol, e.g., less than 0.05% ABV, although the definition differs between countries.

nonreturnable bottle. A glass bottle that is not returned to the brewery for refilling; also referred to as "one-trip bottle."

nucleation. Bubble formation in a wort or beer.

open fermentation. Fermentation in relatively shallow uncovered tanks.

organic acid. Carbon-containing acid, such as citric or acetic acid, released by yeast and largely responsible for the pH fall during fermentation.

organoleptic. Pertaining to taste and smell.

original extract. The amount of extract present in a starting wort as calculated from the amount of non-fermented extract left in a beer (real extract) together with the amount of alcohol produced in a beer. In some locations, e.g., England, it is traditional to describe beer strength in terms of original gravity (OG). For example, a beer may be said to have an OG of 1040. This means that the specific gravity of the starting wort prior to fermentation was 1.040. It is easier to say "ten forty" than "one point zero four"!

oxalic acid. An organic acid found in malt that should be precipitated out in the brewhouse by reacting with calcium to form calcium salt. Otherwise it will precipitate out in beer and beer dispense pipes as "stone."

oxidation. At its simplest, process of deterioration of beer due to ingress of oxygen.

pH. A measure of the acidity/alkalinity of a solution.

pale ale. English-style ale, usually in small pack.

parti-gyle. Dividing the wort collected in the separation stage after mashing into different batches, e.g., the strongest, the mid-strength, and the weakest, to go on to produce different strength beers.

pasteurization. Heat treatment to eliminate microorganisms.

perlite. Volcanic glass used in the filtration of beer as an alternative to kieselguhr.

pesticide. Agent used to protect crops from infection and infestation during growth and storage.

piece. The bed of grain in a malthouse.

pin. A container containing an eighth of a UK barrel.

pint. A measure of beer volume (473 ml in US; 568ml in UK).

pitching. The introduction of yeast into wort prior to fermentation.

Plato. Unit of wort strength (see also balling) that approximates to percent sugar. To convert specific gravity into Plato (to a first approximation), take the two numbers after the zero after the decimal point and divide by 4. E.g., for a specific gravity of 1.040, take the 40 and divide by 4 to yield 10°P.

polyphenol. Organic substance originating in husk of barley and also in hops and which can react with proteins to make them insoluble; also known as tannin.

polysaccharide. Polymer comprising sugar molecules linked together.

polyvinylpolypyrrolidone (PVPP). Agent capable of specifically binding polyphenols and removing them from beer.

pre-isomerized extract. Extract of hops in which the alpha acids have been isomerized.

primings. Sugar preparations added to beer to sweeten it.

propagation. Culturing of yeast from a few cells to the large quantities needed to pitch a fermentation.

protein. Polymer comprising amino acid units.

proteolysis. The breakdown of proteins by proteases.

quality assurance. Approach to quality maintenance that involves establishing robust processes and systems that are designed to yield high-quality product.

quality control. Monitoring of a process to generate information that is used to adjust the process in order to ensure the correct product.

racking. The packaging of beer.

real ale. A term introduced by CAMRA to indicate cask-conditioned ales produced using traditional techniques.

real degree of fermentation. The extent to which the carbohydrates in wort have been converted into alcohol by fermentation. See also **attenuation**.

real extract. The specific gravity of beer after alcohol has been distilled off, this gravity being a measure of the amount of non-fermented material (carbohydrates, protein-derived material) that has survived fermentation.

reduced hop extract. Pre-isomerized extract that has been hydrogenated such that it is no longer light-sensitive and can be used to provide bitterness to beers that are intended for packaging in green or clear glass.

refractometer. Device for measuring the strength of beer.

Reinheitsgebot. The Bavarian purity law of 1516 that decrees that beer can only be produced from malt, hops, yeast, and water.

re-pitching. Practice of taking yeast grown in one fermentation to pitch the next batch of wort.

resin. Fraction from hops that generates the bitterness in beer.

rice hull. The outer layer of rice grains produced during harvesting. Hulls can be used to aid filtration.

rough beer. Beer before filtration.

Saccharomyces cerevisiae. Ale yeast.

Saccharomyces pastorianus. Lager yeast.

scuffing. Scratches on the surface of returnable glass bottles arising from their repeated passage through packaging lines.

seam. The "join" between a beer can and its lid.

shelf life. The period for which beer remains in an acceptable state for consumption.

silica gel. A sand-derived material that can be used to remove proteins that will cause haze in beer.

skimming. Harvesting top-fermenting yeast from the surface of open fermenters.

small beer. Low-alcohol beer made from weak worts, e.g., the later runnings from the mash or historically by the re-mashing of spent grains.

small pack. Cans and bottles.

sour beer. Beer of high acidity (low pH) made using a mixture of microflora, including acid-forming bacteria.

sparging. Spraying the spent grains with hot water during the wort separation process to facilitate extraction of dissolved substances.

sparkler. A nozzle on the end of a dispense tap in the bar that can be loosened or tightened depending on the amount and character of foam expected by the customer.

specification. A parameter measured on a raw material of brewing, on a process stream, or on the finished beer and that must be within defined limits for the material to pass to the next stage in the process.

specific gravity. The weight of a liquid relative to the weight of an equivalent volume of pure water (also referred to as relative density).

spectrophotometer. Device for measuring the amount of light absorbed by a solution at different wavelengths.

spent grains. The solid remains from a mash.

spoilage organism. Microbe capable of contaminating wort or beer.

square. Style of fermenter in that shape.

stabilization. Treatment of beer in order to extend its shelf life.

staling. Deterioration in the flavor of beer.

standard reference method. A procedure for measuring color based on the absorbance of light at 430 nm and described by the ASBC.

starch. Polysaccharide food reserve in barley.

steeping. Increasing the water content of barley by soaking.

stein. German word for "stone," used generally to describe large vessels from which beer is consumed in Germany. Sometimes stoneware, but the term is just as widely used for glass these days.

sterile filtration. Removing microorganisms from beer by fine filtration.

sugar. Small, sweet carbohydrate.

sulfur compound. Flavor active material in beer containing sulfur atom(s).

sweet wort. Wort prior to boiling with hops.

syrup. Concentrated solution of sugars.

taint. Off flavor in beer or a raw material.

tannin. Polyphenol that gives astringent character to beer.

three-glass test. Procedure for blind tasting to discern whether two samples of beer can be differentiated. Also known as triangular taste test.

tintometer. Device consisting of a series of colored discs for comparing with a beer to ascertain whether it has the correct color. Devised by Lovibond.

top fermentation. Fermentations in which the yeast collects at the top of the vessel.

trigeminal sense. Sensation of pain detected by the trigeminal nerve. The mechanism by which carbon dioxide is detected in the mouth.

trub. Insoluble material emerging from wort on heating and cooling.

tunnel pasteurization. Pasteurization of small-pack beer by passage through a heated chamber.

turbidity. Cloudiness (haze) in beer.

ultrafiltration. Filtering out of material at the molecular level by passage through very fine membranes.

viability. Measure of how alive something is.

vicinal diketones. Butterscotch- or popcorn-flavored compounds formed during brewery fermentation.

volatile. Describes a molecule in beer that contributes to aroma and is easily driven off.

wet hopping. Use of undried, newly harvested hops to flavor beer. Also known as fresh hopping.

whirlpool. Vessel for separating trub from boiled wort.

widget. A device in a can (sometimes bottle) that promotes the nucleation of foam.

wild yeast. Any yeast that is not associated with a particular beer.

wort. Fermentation feedstock produced in the brewery.

wort separation. Act of separating sweet wort from spent grains.

xerogel. A type of silica gel.

yeast. Living eukaryotic organism capable of alcoholic fermentations.

zentner. 50 kilograms of hops.

zinc. A metal that promotes yeast activity and foam stability.

zymurgy. The study of fermentations by yeast.

Index

For the benefit of digital users, indexed terms that span two pages (e.g., 52–53) may, on occasion, appear on only one of those pages.